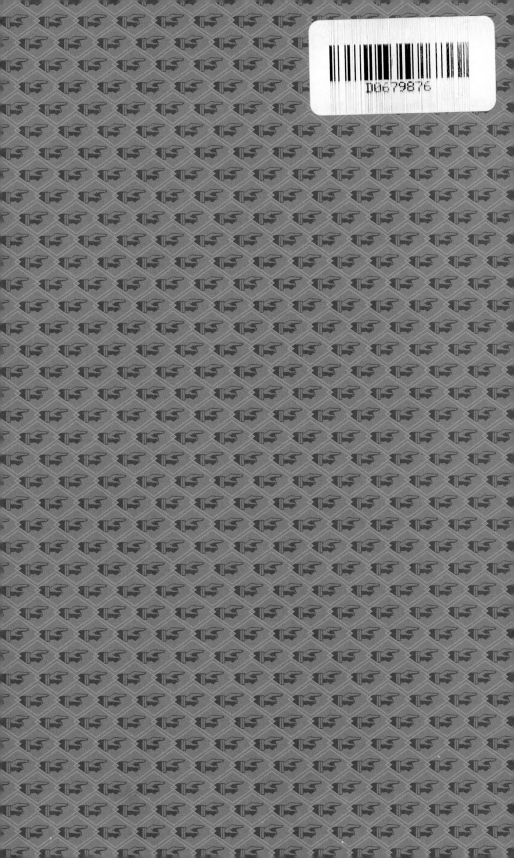

A HISTORY OF
THE WORLD
THROUGH BODY PARTS

Petras & Petras

A

HISTORY

OF THE

World

⇥ THROUGH ⇤

BODY PARTS

The Stories Behind the Organs,
Appendages, Digits,
and the Like Attached to
(or Detached from)
Famous Bodies

KATHRYN PETRAS

&

ROSS PETRAS

CHRONICLE BOOKS
SAN FRANCISCO

TABLE OF CONTENTS KEY

CONTENTS

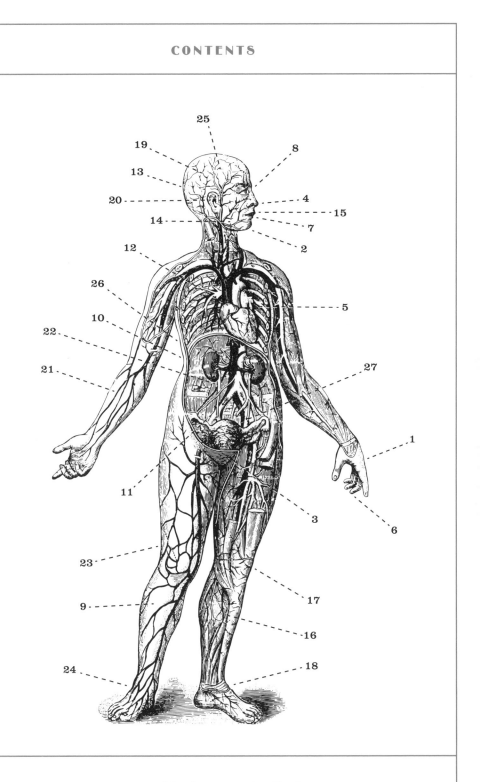

DIAG. Table of Contents Visualized.

TABLE OF CONTENTS

INTRODUCTION

HE IDEA FOR THIS BOOK came from a famous nose. Specifically that of Cleopatra—and more specifically from mathematician Blaise Pascal's famous observation about it:

"Cleopatra's nose, had it been shorter, the whole face of the world would have been changed."

Pascal's nasal focus was philosophical: Cleopatra's nose was, in his eyes, trifling in one sense, but hugely important to world history in another, because it captivated Julius Caesar and Marc Antony and through them greatly affected the West's greatest empire. Pascal's speculation has gone on to spawn numerous discussions of contingency theory and how important seemingly minor events such as the size of a nose can be to later events in world history.

But our interest was far more prosaic. Forget the contingencies of later world history, we said. What about that nose? Was Cleopatra's famous nose actually that long and captivating? And secondly: If so, how did Pascal know this? And thirdly: Why did it captivate Caesar and Antony? Particularly in our age of ubiquitous rhinoplasty, this seemed, well, a different cultural desideratum. What was the Roman attitude toward noses and why were they such a big thing? Or were they?

We did a little research, and, in so doing, became fascinated by body parts and their role in history in general, and their reflections on the societies in which they existed specifically. In short, we found that one can learn a great deal by focusing, as did Caesar and Antony, allegedly, on an individual body part. And so began our journey into discrete pieces of history (and discrete pieces of bodies). Beginning with Cleopatra's nose, we became captivated by more and more historical body parts—from famous craniums all the way down to infamous feet, from prominent breasts to bygone bowels. Most important, we realized what so many of us miss when reading or thinking about history: the human body. Yet of course we all have one, and so did all of those figures in history. So why do we so frequently ignore the body?

In this book, we look at different body parts in history—specific famous or infamous body parts of specific historical figures, more general body parts as related to specific cultures and ideas of the times—presented in chronological order, from paleolithic hands to space-age bladders. We also address questions such as, What did the people of the past feel about their bodies? What did they do with them? What part did they play in history? And how can we understand their lives and culture more by looking at their certain telling body parts?

We found that zeroing in on a body part can lead to fresh and often surprising insight into ideology or thought. Through looking at Lenin's moldy skin and the mechanics of body preservation, we see Soviet communism more as an extension of medieval religion (and specifically, in Lenin's posthumous case, as textbook hagiography) than as a "modern" political economic system. Through ancient Egyptian ruler Hatshepsut's beard and Vietnamese heroine Bà Triệu's breasts, we see the power of the patriarchy and the struggle even the most prominent women faced. In short, smallish body parts give us a biggish picture of the human condition.

Like it or not, we are all embodied beings with all the problems and glories of fleshly, bodily existence. We all have functioning, and sometimes partly or nonfunctioning, body parts, and they play a role in our lives and in our thought. In some cases, they may actually direct our thought, although causation is hard to prove. Would Tamerlane have been as monstrous with two fully functioning legs? We can only speculate. Would the Reformation even have occurred if Martin Luther had fully functioning bowels? We don't know. But we do know that Luther frequently alluded to his chronic constipation and admitted that he thought of his famous 95 theses in the "cloaca," Latin for sewer, what is thought to have been Luther's euphemism for toilet.

Indeed, our bodies, as we all know, have their unpleasant or little-talked-about aspects. And that is what makes bodily history so interesting—it truly is *human*, warts, bowels, noses, and all. We can learn much from the historical body, although it has been so overlooked.

By focusing on body parts, we've tried to make history truly human in ways one might not expect and make people from the past come alive. Take Martin Luther and his bowels. Based on our research, we now think that the pained, strained expression he commonly wears in paintings and engravings is suggestive in ways that are perhaps unwarranted but certainly plausible! And certainly and more seriously, each body part we cover is a jumping-off point for a wider look at the times.

A
HISTORY
OF THE
World
⊰ THROUGH ⊱
BODY PARTS

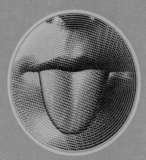

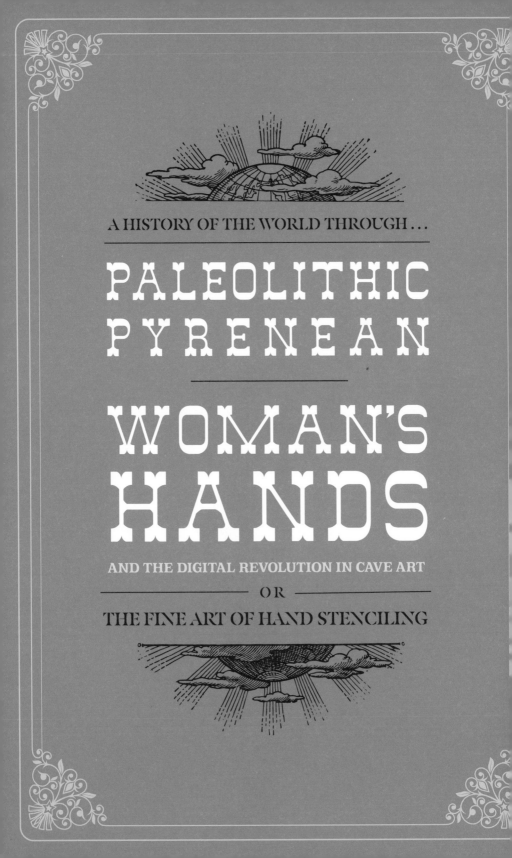

A HISTORY OF THE WORLD THROUGH...

PALEOLITHIC PYRENEAN

WOMAN'S HANDS

AND THE DIGITAL REVOLUTION IN CAVE ART

—————— OR ——————

THE FINE ART OF HAND STENCILING

(50,000–10,000 BCE)

HAT WAS THE WORLD'S FIRST art movement? Think hands. As in hand stenciling,
simple outlines of hands usually in either red
or black. This was The Happening Art Thing
beginning 40,000 or more years ago. Strange prehistoric hand stencils
have been found on cliff faces and particularly on walls deep in caves
all from Argentina to the Sahara. They're probably the first human
artworks, the first time humans interacted with the environment to
make something other than a utilitarian rock tool. And they were,
in effect, a major art movement—a hand-art fixation that lasted for
tens of thousands of years, which is a lot longer of a creative run than
impressionism or pop art. So what are these hands trying to tell us?

Let's first take a trip back in time 28,000 years or so and go deep
inside the limestone caves of Gargas in the French Pyrenees mountains,
and tag along with some putative "cavepeople" artists so we can see how
it was done (we think). We're with five others, a young woman, a young
man, two teens, and a seven-year-old. Back then, deep cave ventures
were usually done in multigenerational groups, probably families.
After walking barefoot for almost half a mile in mostly darkness and
almost total stillness punctuated only by the sound of water dripping

Hand Stencils Aren't Dated, So How Do We Know How Old They Are?

TO DETERMINE THE AGE OF HAND STENCILS, SCIENTISTS USE uranium or uranium-thorium dating, usually by analyzing mineral formations overlying the art. Caves tend to be wet; over the years, water and dissolved minerals (carbonate or calcite, along with trace amounts of radioactive uranium and thorium) drip over the stencils from the walls and ceilings and form a residue. Meanwhile, the uranium in this residue is naturally decaying into thorium. Scientists take samples of the residue, analyze the ratio of uranium to thorium, and extrapolate backward to determine age—the more thorium in the lowest layer covering the artwork, the older it is. There are complications to this, of course; minerals can flake off, chemical reactions can occur, and the scientific protocols of measuring can vary. But with some refinements, this kind of dating can give a fairly good idea of age. Back in 2016, a hand stencil in a cave in Borneo, Indonesia, was dated back to a maximum of 51,800 years, a world record. Humans have been hand stenciling for a long, long time.

from stalactites, and after sometimes ducking through narrow and low corridors with clay floors, we reach a large gallery. Someone lifts up their light source, cleverly made of several resinous pine sticks bundled together which allows decent illumination, and we see 200 hand stencils all outlined in either red or black. It looks something like a weird surrealistic flower garden growing on the cave wall. Roughly half of the hand outlines are macabrely missing parts of fingers as if they've been chopped off. Now the woman (most handprints are feminine) raises her hand and places it against the cave wall. Then either she or someone else puts some red pigment (either dry, to be mixed with spit or water, or premixed with bear fat) into their mouth, aims a hollow bone tube at the hand, blows, and voilà!—as their French will say 30,000 years later—a hand stencil is formed.

The thing that has perplexed anthropologists is why. Why go to all the trouble of venturing half a mile into a cave to put up a picture of a hand? One idea is the "palpation theory"—the hand stencils serve as guides or signs for others entering deep into the cave: Watch out, stop, go left. But this doesn't explain *why* they're trying to go into the cave in the first place. A more integrative theory combines both the stencils and the cave venturing—the cave is the entrance to the deep underworld and the hand stencils are shamanic ways of touching the spirit world, seeking to enter it and receive favors, with the cave walls representing Mother Earth. Then there is the simple "I'm here" theory—hand stencils as prehistoric graffiti or art—i.e., caveman Banksy doing his stuff.

Interestingly, it appears that hand stenciling has persisted among indigenous Australians from Paleolithic times to the present, spanning the 50,000 years of development of their complex society. Some 20th-century researchers interviewed practitioners to see how this age-old art was manifested in modern times. An unexpected finding: The Aboriginal people could recognize the handprints of their relatives. Hand stencils were literal "records of individuality," personal signatures, often with a religious twist. As one researcher put it:

It is the belief of a native of the north-west that the spirits of departed tribes-people desire to be revered by those nearest to them; and for that reason they keep a tally of their visits made to the sepulchral caves. By placing the imprint of his hand upon the wall, the native leaves evidence of his call . . . Each hand-mark can be recognized . . . by every member of the tribe, with wonderful precision and reliability. (Basedow 1935)

That's a high degree of sight-sophistication, not surprising in a culture accustomed to tracking animals, but it's hard for us to imagine. Try differentiating handprints from 50 people yourself. Perhaps hand stencils back in prehistory, then, were ownership tokens too. Australian researchers also discovered a possible reason for the "chopped" or attenuated finger stencils found in a few caves, including the Gargas caves. Early French researchers thought that the women making the hand stencils literally (and rather bizarrely) were missing digits. But the Australians showed otherwise. By careful placement of bent fingers, stencils could be made that looked like (but weren't) hands with parts missing, representing bent-finger hand signals. The Aboriginals traditionally had a sophisticated signing vocabulary—very valuable silent talk for hunters stalking prey and probably also used by Paleolithic people who didn't want to alert some random hungry cave lion.

One thing we're fairly sure about: Hand stenciling came before virtually all other forms of human art communication. The earliest hand stencils were done about 45,000 years ago and were probably not even done by humans as we know them but by our unjustly belittled beetle-browed cousins, the Neanderthals (whose brain capacity, incidentally, was actually about 10 percent larger than ours). The "high cave art" period—with the beautiful prehistoric paintings of horses, deer, and cave bears—didn't come until thousands of years later.

So for a good 10,000-year run, hands were pretty much It. And not just hand stenciling. There are also the minor hand arts, as it were, such as finger fluting (lines left by fingers on cave clay) and hand,

A Cave Art Caveat

JUST TO MAKE THINGS COMPLICATED, RECENTLY, IN SUMATRA, Indonesia, a cave drawing, not of a hand but of an unknown animal, was discovered and dated back (possibly) an unprecedented 50,000 years, drawn by one of the world's first modern artists.

palm, and thumb printing. Even the other big art thing, dots, often decorated handprints and stencils. But why hands?

Well, the hand is iconically human. It let us make sophisticated tools, manipulate the environment, and beat out competitors such as cave bears and lions. The thinking now is that the human brain evolved in tandem with the hand. Manual gestures pushed brain evolution. Hand movement may have been the primary activator of human cognition and communication ability, while sophisticated speech came later. A large chunk of the brain is involved with the hand too; it takes a lot of brain to figure out how to manipulate one.

On the peculiar art side of things, hand stenciling or printing was also one of the easiest ways of actually creating art, of coming up with a representation of something that was real and utterly human. This in turn gave people the feeling that they had a measure of control over an often hostile environment.

It took a lot longer, around 10,000 years, to figure out how to draw a horse.

Finally, on the spiritual side of things, as mentioned before, hand stencils may have been a way for humans to communicate with the spirit world. Religion and art have been intertwined since history began. Therefore, hand stenciling might not only be the first art movement, but also the first *religious* art movement—a lot earlier and more primitive than the sophisticated ancient Egyptian tomb paintings of humanoid goddesses like Isis, or the Renaissance depictions of Jesus and Mary, but, hey, you gotta start somewhere.

Ultimately, whatever the reason, there's no doubt that hand stenciling is, as one scholar put it, "the earliest known artistic symbol of the human form"—humanity's first self-portrait.

A Short Trip Back 28,000 Years

THE LITTLE CAVE EXCURSION WE MENTIONED EARLIER IS BASED on fact—and an amalgamation of studies of humans venturing into caves. The first investigation at the Bàsura cave in Italy found traces of five humans traveling a half mile underground. Scientists calculated the age and sex of the people by analyzing hardened footprints, knee prints, and finger and hand marks in cave clay deposits. The second investigation involves another cave, this time in Spain, where scientists carefully measured angles of hand placement and hand-digital ratios and concluded that most of the handprints were from women, were mostly left hands, were mostly stenciled at shoulder level, and that someone else did the actual stenciling probably using a hollow tube to blow red (ochre) or black (manganese oxide) pigment. Hollow bones have been found with red pigment in them, lending credence to this. The pigments were said to have been made using a mixture of iron oxide and manganese crushed with animal fat. The third investigation was from the Gargas caves in the Pyrenees mountains in France, which is famous for its deformed handprints, and which is where our party of five entered that cave one day 28,000 years ago.

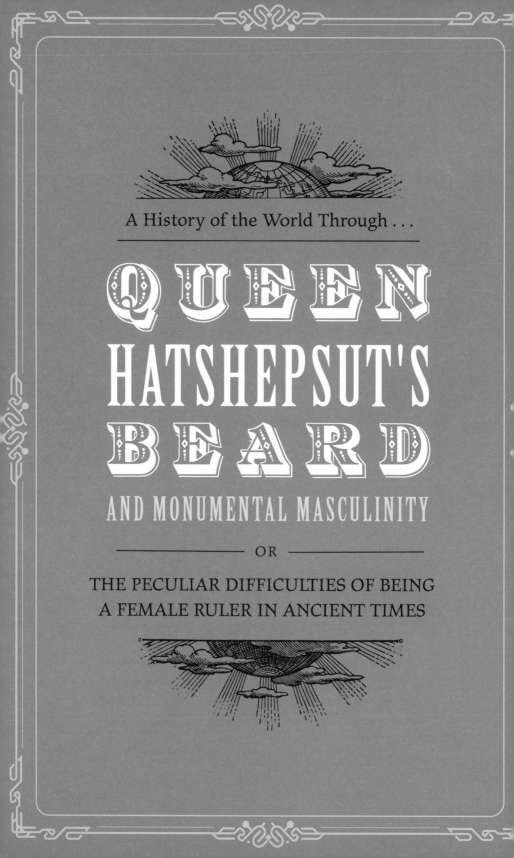

A History of the World Through . . .

QUEEN HATSHEPSUT'S BEARD

AND MONUMENTAL MASCULINITY

— OR —

THE PECULIAR DIFFICULTIES OF BEING A FEMALE RULER IN ANCIENT TIMES

(FL. 1450 BCE)

HEN THE FIRST TRANSLATOR OF
ancient Egyptian hieroglyphs, Jean-François
Champollion, visited the colonnaded ancient
Egyptian ruins at Deir el-Bahri, he noticed
something a little . . . unexpected: a bearded lady.

> Still more astonished was I to find upon reading the inscriptions
> that wherever they referred to this bearded king in the usual dress
> of the pharaohs, nouns and verbs were in the feminine, as though a
> queen were in question. I found the same peculiarity everywhere . . .

This king, dressed as a man but addressed as a woman, was Hatshepsut, fifth pharaoh of the Eighteenth Dynasty of Egypt, who took
to the throne—and put on the beard—in 1478 BCE. Wearing the beard
was no idle whim. A pharaoh's beard was the ultimate ancient Egyptian power symbol. Wearing one exemplified her political power and
legitimacy, which typically were vested in male bodies.

As you might have guessed, Hatshepsut's beard was fake—probably metal with goat hair attached to a gold chin strap, with cords on
both sides extending up to the ears to keep it in place—but not just

because she was a woman. Even male pharaohs wore artificial beards in depictions and during ceremonies.

Real facial hair was a fashion faux pas at the height of ancient Egyptian civilization. Egyptians—including pharaohs—were clean-shaven in daily life. In tomb and temple paintings, typically the only beards depicted (other than the pharaohs' fake ones) were on captured enemies. There were some exceptions to the clean-shaven rule in daily life, but usually a good Egyptian man had a good clean face. Priests took it one step further and went for the full manscaping, apparently shaving off all of their body hair as well. Only the poorest of the poor seemed to forgo shaving.

So why the false beard on pharaohs? No one is sure exactly why they wore them, but the beard was connected with the gods, particularly the great risen-from-the-dead Osiris, lord of the underworld and judge of the dead, who was always depicted with a beard, and not just any beard but one that looked like it was fake itself. The artificial beard underscored the link between the ruler and the eternally reigning god and reinforced the idea that the pharaoh was divine as well. No wonder, then, that Hatshepsut opted for the hairy strap-on like the men before her. Even more than they, she needed to broadcast her status as divine ruler.

Despite having a uterus and other biological female accoutrements, Hatshepsut met the key (and really only) job requirement of being pharaoh: In other words, she was the child of a pharaoh, Thutmose I. When he died, she married his son, Thutmose II. He wasn't her brother, as you may have thought, just her half-brother. And he wasn't a terribly effective pharaoh; in fact, Hatshepsut supposedly ruled from behind the throne. When he died, the throne went to her nephew, his son from a minor queen, cleverly called Thutmose III. III was only two, so Hatshepsut was named queen regent, his ostensible co-ruler. A two-year-old child isn't typically that practiced at statecraft, so yet again Hatshepsut really called the shots.

| The Mummy's Molar

AS WITH SO MANY OTHER PHARAOHS, HATSHEPSUT'S MUMMY wasn't found in her tomb. So where was it? In 1903, Egyptologist Howard Carter (of King Tut fame) discovered a tomb called KV 60 (Valley of the Kings Tomb 60) that held two female mummies. One had an inscription identifying her as Sitre-In, Hatshepsut's wet nurse; the other was not identified. And that was that until 2007, when archeologists examining the unidentified mummy noticed it was missing a tooth. Not a big deal, except for the fact that, in another tomb near Hatshepsut's great temple, archeologists had found a canopic box—a burial container containing body parts—with Hatshepsut's royal name on it. Inside was a decayed mass of liver and a single molar tooth, which almost exactly fit the space in the unknown mummy. Was the mummy Hatshepsut, then? So far, DNA tests matching it with other related Thutmosid mummies are inconclusive, but if it is, the body is quite different from the idealized statuary figures of her, both male and female. A CT scan revealed an obese woman, aged between 45 and 60, with bad teeth and most likely type 2 diabetes. Sadly, she also seemed to have had bone cancer, which is speculated to have been caused by liberal use of a traditional skin salve that might have moisturized dry skin but also had an unexpected side effect—it was carcinogenic.

But after seven years as queen regent, she wanted the real power and the actual role of a full-fledged pharaoh. She adopted the formal titles and duties and, to make sure her rule appeared legitimate, stressed her lineage. Like any good political operative, she made sure that everybody who was anybody knew that her father had actually wanted her, not her brother/husband, to be heir.

Then, like the practiced politician she was, she took it one step further and launched the ancient version of an image-identity campaign—starring herself as a gender-bending pharaoh. The "male" version of Hatshepsut evolved as she morphed from queen regent to actual pharaoh. At the beginning of her reign, there were just hints of masculinity: Statues showed her with female features but with masculine accoutrements such as a khat headcloth or, yes, a beard. Later she appeared androgynous, as in a life-size limestone statue from her Temple of a Million Years—she is topless as a typical male would be, has only a slight hint of female breasts, but has slim shoulders, delicate facial features, and no beard. It's far from the traditional clearly identifiable male or female image.

The changes continued, and the masculinity became even more obvious. Take a look at one of her later statues, and she has become a full-fledged man with wide shoulders, no hint of breasts but instead firm pecs, masculine facial features, and the wide flared beard of a pharaoh. Her skin color as depicted on tomb walls changed as well—from the traditional yellow of females to an odd mixture of yellow and red and finally to the ruddy red used for Egyptian men, presumably tanned by warring, hunting, or fishing in the African sun.

For all the male imagery in her public persona, Hatshepsut's masculinization was probably your textbook political ploy, not a personal gender-identification issue. She apparently continued to identify as female in her personal life, stuck with female gender pronouns in temple inscriptions, and even tried feminizing some of the traditional pharaonic titles. And possibly for protection against misogynist males who resented her power, she put the powerful Theban goddess Wosret and the cobra goddess Wadjet in her royal names along with Ma'at, the goddess of the cosmic order. This underscores one of her guiding rules of kingship: The best offense is a good defense (especially against envious men). And it worked. From all reports, she achieved a great deal as pharaoh. She advanced Egyptian trade, oversaw numerous building projects (including her famous mortuary temple, the Temple of a Million Years, where she first became known to the modern world), and is generally considered to be one of the most successful pharaohs, male or female.

There's tantalizing evidence that she attempted to raise her daughter Neferure to her kingly status to perpetuate her lineage and the revolutionary idea of a female pharaonic line, but that attempt failed. In fact, it took another 1,200 years, give or take, until another woman became ruler (and goddess). Arsinoe II married her younger brother Ptolemy II (still keeping it all in the family) and, as co-ruler, took the throne—but not the beard. (She was mostly of Greek-Macedonian origin, and unlike the Egyptians, it just wasn't the thing for them.)

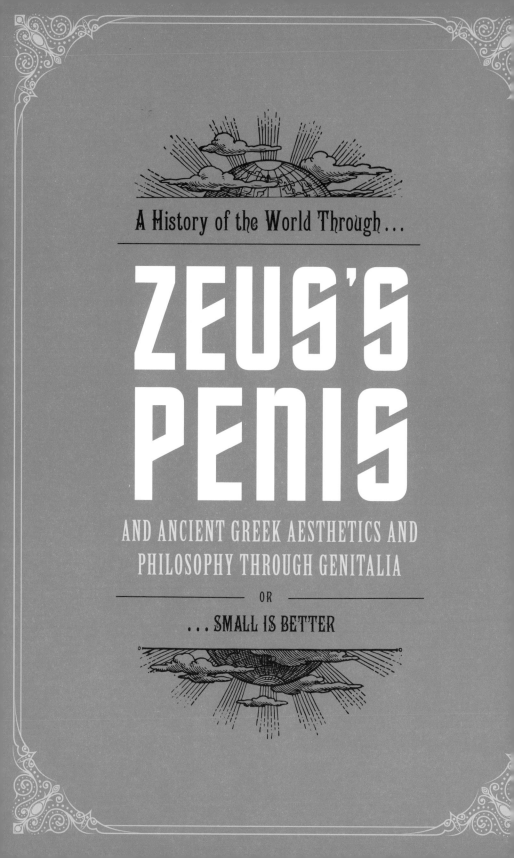

A History of the World Through...

ZEUS'S PENIS

AND ANCIENT GREEK AESTHETICS AND PHILOSOPHY THROUGH GENITALIA

OR

...SMALL IS BETTER

I F YOU LOOK AT A FAMOUS ancient Greek statue, the *Artemision Bronze*, for example, which depicts mighty king of the gods Zeus, you'll notice something glaringly obvious or, actually, glaringly *not* so obvious. Zeus's penis. It's . . . tiny. And we're not talking shrinkage.

We can't single out just Zeus. Virtually all statues of Greek heroes and gods sport similarly piquant petite penises. Clearly ancient Greece was very different from the modern United States when it came to what constitutes an enviable male member, as is made clear in this soliloquy summing up the masculine ideal from comic playwright Aristophanes's *The Clouds*:

> If you do these things I tell you, and bend your efforts to them, you will always have a shining breast, a bright skin, big shoulders, a minute tongue, a big rump and a small prick. But if you follow the practices of today, for a start you'll have a pale skin, small shoulders, a skinny chest, a big tongue, a small rump, a big prick and a long-winded decree.

Yes, Aristophanes and his fellow Greeks believed a "small prick" was a characteristic of the perfect male. This paragon also had a tanned, lightly muscled, athletic body, and a calm, deliberate mind—in sum, the ideal embodied in the ancient Greek word *sophrosyne* (σωφροσύνη), or excellence of character.

Sophrosyne was everywhere in ancient Greece: in plays, in philosophy, and in daily life, where it was what a man was expected to strive for. (Women—lustful, inferior beings who were basically expected to stay at home—were pretty much out of the picture.) To a classical Greek, a large, bulging penis simply didn't fit with sophrosyne. It was—literally and figuratively—too out there. Classical Greek statues reflected the tempered sophrosyne ideal with not a muscle out of place and nothing bulging excessively.

Small penises also represented sophrosyne in another way. Aesthetically and philosophically, big penises were bad. So were erect ones. They stood for out-of-control, mindless sex, which was definitely not sophrosyne. There was a proper place, though, for the large and/ or erect penises in art—on ithyphallic (straight penis) statues, in the restrained parlance of museum curators. These statues still are often tucked in the back galleries of museums. In fact, until the 1960s, the British Museum kept them all in the Secretum, a secret cupboard for "rude" objects, completely away from the "innocent" public. The statues sport huge penises usually attached to grotesque gods such as goat-footed Pan or the dwarf Priapus (who gave his name to the medical condition of a perpetual erection: priapism). And that's the point: The Big Penis Gods are far from a human ideal, and far from the Small Penis Gods one looked up to.

There's a side note to all this, a double standard of sorts that speaks of male privilege. On one hand, you have small and dainty polite public penises, while on the other hand, in erotic art meant for private enjoyment, you find jaunty, bulging penises ready for action. Men got the best of both worlds: sophrosyne-imbued dicklets in public, raging bulls in private.

The Problem with (Classic) Penises in Literature

ARISTOPHANES'S PENCHANT FOR PENISES AND OTHER EROTIC parts created problems for priggish Victorian English and American translators. His πέος (*peos*) or *cock/prick* was changed to an oh-so-discreet and cloying "the joys of Love" in one famous 1878 translation of his play *Lysistrata*; other times, the offending parts were simply removed.

The Science of Classical Statues

CLASSICAL GREEK STATUES WEREN'T JUST EYEBALLED depictions of people, they were mathematically derived. The last joint of the little finger was the basic module for determining proportions—scaling everything up by the square root of 2 to get other measurements of fingers and the arms; and from there everything else was scaled, all the way down to that quintessential scale model of a little penis. And those small penises also had science behind them. None other than Aristotle, the father of Western science, noted in his *Generation of Animals* that a large penis is linked with male infertility, and no Greek male wanted to be accused of that.

Why a hierarchy of penises at all? Nowadays we don't generally find people focusing on the ideal penis. Sure, size is an issue in porn and even in genteel erotica with a general idea that bigger (to a point) is better, but except for penis enlargement ads and the "take a look at my junk" sext, it doesn't figure too much in daily life, particularly in the aesthetic sense. By contrast, ancient Greeks had penises on their minds. People thought a lot about what the best penis would look like, with large foreskins particularly admired. Greek nurses used swaddling to elongate the foreskin of babies and to pleasingly shape the scrotum. Circumcision was a no-no, something only for enslaved people who were often shown on vases with huge circumcised penises. Circumcision was also a real knee-slapper in comic plays, particularly as a source of anxiety.

Maybe one reason for all this attention to all things penis was their ubiquity. Your average Greek males were literally surrounded by a plethora of penises almost from birth. In their early years, they trained athletically in a gymnasium in the nude. (*Gymnos* means "nude" in ancient Greek; the verb *gymansion*, "to exercise naked.") Walking around town, they'd see nude statues of gods and athletic heroes with genitals painted in living color at temples, marketplaces, and other public areas. At many street corners, they'd see herms—plinths with a sculpted head on top and a penis and scrotum sculpted into the base where it would be on a man. (These look quite bizarre to modern eyes and aren't seen much nowadays. One reason: Often the penis-adorned lower part of the plinth is cropped out in art history textbooks so we see only the dignified head.) Greeks would walk past public buildings decorated with reliefs of heroes, gods, and men in battle, again often nude with genitals clearly on display. At home, they'd drink their wine from cups decorated with nude buff males doing their thing. (Interestingly, Greek vases more often depict heterosexual intercourse as well as larger erections.) Watching plays, they'd see actors wearing artificial phalli made of thick red leather. These usually hung down, but if the need . . . arose, special erect phalli were used as well. And on feast days, your basic ritual procession wouldn't be complete without people carrying a huge model penis or two.

Clearly there was much ado about male genitalia in ancient Greece, especially as depicted on statues of gods. But what about women? What about the statues of the goddesses? Here's an even odder aspect of Greek art: At your next museum visit, take a close look at a corresponding statue of an ancient Greek goddess. You'll notice something that is not small but completely missing: Classical Greek statues of women have no genitals whatsoever, no protruding labia, no pelvic mounds, no vagina, nothing at all to suggest sexuality or sex. Why? Many scholars believe it's a misogynist (Greek word!) extension of the ideal of the small penis. Men were supposed to be controlling their urges, while women, who were seen as sexually aggressive, were denied the opportunity to even try . . . at least in statuary.

Sausages, Knobs, and Ancient Greek Penises

FOR ALL THE HUBBUB AND FASCINATION THE ANCIENT GREEKS bestowed upon the penis, it wasn't treated as a holy object. Like us, the ancient Greeks had obscene and slang terms for the organ, many of which are quite familiar. Among them: *sausage, meat, loaf, tender horn, crow, eels, knob, so-and-so, finger, shaft, goad, point, horse, ram, bull,* and *dog.* A "beam man" was a man with a large penis; a "mushroom" was a hard-on. The word *peos* meant "cock" or "prick." The polite term was *phallos, posthe* (an affectionate term for a boy's little penis), or *sathe.*

The Exception
to the Rule

IN THE REALMS OF PENIS ARCHEOLOGY, THERE'S A RECENTLY discovered exception—a bona fide hero with a large, erect penis. It's a small 8th-century BCE bronze sculpture (very early in Greek history), the earliest surviving depiction of Ajax, a Greek hero of the semi-mythical Trojan war. But this isn't your typical Ajax. The 3-inch statue shows Ajax with a large hard-on. Moreover, the statue as a whole looks like a larger representation of a hard-on—so it's a penis within a penis. Why would a hero be so grotesquely (to a Greek) phallically large? One conjecture: The statue depicts Ajax as he is about to kill himself with his sword, because he was enraged that Odysseus rather than he was rewarded with dead Achilles' armor. This rage is a dishonorable thing, so perhaps the large penis is a sign of disgrace, a penis that befits a man who has lost his sophrosyne.

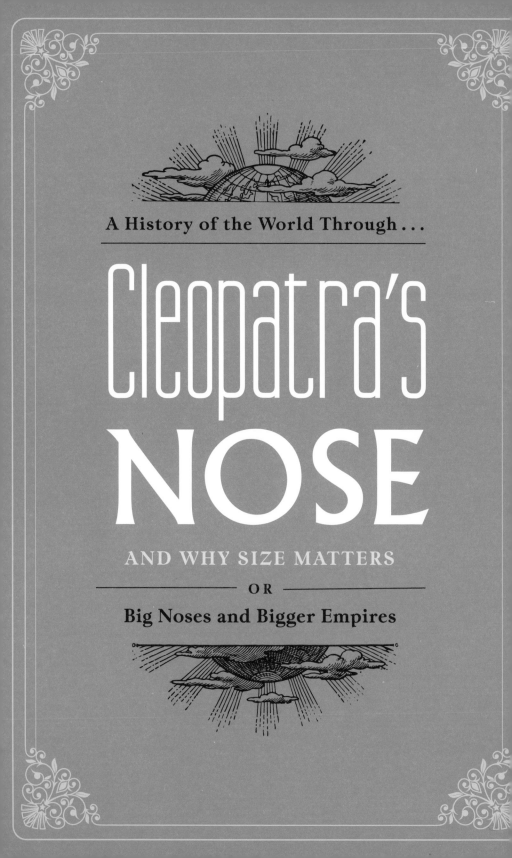

A History of the World Through...

Cleopatra's
NOSE

AND WHY SIZE MATTERS

—————————— OR ——————————

Big Noses and Bigger Empires

(69–30 BCE)

 CCORDING TO 17TH-CENTURY French philosopher Blaise Pascal (also a member of the larger-than-average-proboscis club), size mattered in a very *big* way, and so he came out with his famous aphorism: "Cleopatra's nose, had it been shorter, the whole face of the world would have been changed."

And so began centuries of philosophical inquiry—and more than a few forays into historical detective work—all to ascertain (1) would the world really have been different if Cleo's nose had been smaller, and (2) was it really that long anyway?

To address the first question, a bit of history: Cleopatra (69–30 BCE) was the queen of Egypt just as the old Roman Republic was being transformed into the Roman Empire and looking hungrily at other nations to conquer. Cleopatra had successfully seized the throne from her brother with foreign aid courtesy of Roman dictator Julius Caesar (whom, by all accounts, she captivated, even bearing him a son). But when he was assassinated by Romans who were threatened by his growing power, she worried about protecting her throne and Egypt's independence. Her move? To captivate (and ally with) one of Caesar's subordinates who was jockeying for his position. But she chose the

Caesar's Comb-Over and Cleopatra

JULIUS CAESAR, DICTATOR AND ALL-POWERFUL LEADER OF ROME (until the Ides of March, that is, when he lost all power and his life), had it all—except a full head of hair. He was very self-conscious about his receding hairline. His battlefield enemies supposedly wouldn't stop taunting him about it, so, like certain other modern leaders, he sported a characteristic and rather pitiful comb-over. All to no avail. The Roman patrician was balding badly, and a recently discovered statue, found in modern-day France, shows what his enemies were talking about: a definitely receding hairline not disguised at all by those carefully combed-over wispy strands of hair. This leads us to speculate: Did Cleopatra offer her Roman lover some of her anti-balding medications? Although some scholars doubt it, she was said to be the author of Cosmetica, an early treatise not merely on cosmetics, but also on dermatology. Her anti-balding medications, as reported by the later Roman doctor Galen, were not a bit like Rogaine: "The following is the best of all . . . for people getting bald all over. It is wonderful. Of domestic mice burnt, one part; of vine rag burnt, one part; of horse's teeth burnt, one part; of bear's grease one; of deer's marrow one; of reed bark one. To be pounded when dry, and mixed with plenty of honey till it gets the consistency of honey; then the bear's grease and marrow to be mixed (when melted), the medicine to be put in a brass flask, and the bald part rubbed till it sprouts."

wrong guy, Marc Antony. They were ultimately defeated by another faction led by Caesar's nephew, Octavian (later Augustus Caesar). Cleopatra apparently tried the Captivate-a-Roman-Leader move again, but Octavian wasn't buying. After he conquered Egypt, he instead politely suggested that she commit suicide.

An important factor in all this political jockeying as far as Cleopatra was concerned was, according to Pascal, her nose. Many leaders of Rome happened to proudly sport long so-called aquiline noses with a hook at the end—and also happened to come up with the convenient idea that long noses signaled dominant personalities. Enter Pascal's theory: Without a good long nose, Cleopatra wouldn't have fascinated long-nosed powerful Romans like Julius Caesar and Marc Antony—and thus Rome, and the entire Western world, would have turned out quite differently.

This idea is actually an early iteration of what today is called chaos and/or contingency theory—the idea that a small event or thing (i.e., a nose) may trigger immense changes down the road. And so along those philosophical lines we have myriad modern scientific papers with titles such as "Cleopatra's nose and the diagrammatic approach to flow modeling in random porous media" dealing with such issues of small non-nasal perturbations causing much larger ones.

The second question—how long was Cleopatra's nose or, for that matter, was it long at all?—is much harder to answer. The main problem is that ultimately Cleopatra lost her Egyptian empire to Rome while quite young. Losers tend not to write history books, and their portraits, statues, and coins tend to be destroyed or not displayed. There are one or maybe two or maybe even twenty statues of her, depending on who's doing the interpreting. Problem is, none of the statues have a handy "Cleopatra VII" etched on them, so there's a lot of conjecture. A marble head now in the Vatican is the only one virtually all scholars agree is definitely Cleopatra, based on her melon hairstyle with a bun and a royal diadem. But it falls short when it comes to olfactory deduction: As with many ancient statues, the protruding nose has chipped off. A number of other more problematic statues are also sans nose.

For more authoritative nasal detective work, scholars have turned to the handful of coins that have been found. There's one coin in particular, a bronze coin minted in her capital city of Alexandria, that seems quite definitive. She appears, according to one scholar, quite "radiant" . . . and also with a long, rather large nose. Later coins that Cleopatra had minted along with Marc Antony are not so flattering. Her nose is still long but her neck and features are rather massive. She looks more like a female counterpart of fleshy Marc Antony or a strong contender for the ancient equivalent of the WWE than a woman who sexually and emotionally captivated Rome's elite.

So now we have a new problem: Did Cleopatra change or did her portraits change? She was undoubtedly petite (she was once easily smuggled in to see Caesar, hidden in a rug), so the best guess is that she changed her portrait to match Antony's as a political act—to show her equivalence to a powerful Roman. (In fact, she one-upped Antony. On several of their joint coins, she's in front, the obverse, and he's on the back.)

Speaking of changing depictions, there's yet another wrinkle in the "What did Cleopatra look like?" saga. Although entirely or mostly ethnically Greek (there's some debate about her maternal grandmother), Cleopatra, as queen of Egypt, was considered divine, a god. As any good Egyptian god would, she built temples and put carvings of herself on the walls in traditional Egyptian attire, all with a typically stylized short Egyptian nose.

So was Cleopatra's nose actually all that long? Probably yes, along the lines of the depictions showing a biggish but thinnish nose with a slight crook, much like the best of Roman aquiline noses. But Cleopatra probably did some growing and trimming of her nasal portrayals depending on context, diminishing it for the traditional Egyptian portraiture and enlarging it (and her neck) when in battle for her life with Marc Antony against Octavian.

It was all part of being an intelligent leader, which Cleopatra— with any nose—surely was. Did we mention she spoke nine languages (including Aramaic, ancient Egyptian, Greek, and Latin) and may have written learned treatises on dermatology, while keeping her day job as the ruler of Egypt . . . which leads us to think that perhaps Pascal was wrong. Even without that long nose, we're quite sure she would have done it all anyway.

A Wall of Noses

ONE PROBLEM WITH MOST ANCIENT STATUES LIKE THOSE OF Cleopatra is that protruding parts, particularly noses, get broken off over time, and very often not by accident. In fact, according to Professor Alex Mullen of the University of Nottingham, cutting off marble—and, for that matter, human—noses was a bit of a thing in ancient times. Egypt even had a settlement called Rhinokolura (ancient Greek for "docked noses") where criminals whose noses were sliced off as punishment were sent; the deposed Greek Byzantine emperor Justinian II had his sliced off as well, and a statue of Tiberius Caesar's famous nephew Germanicus shows evidence that his basalt nose was clearly chipped off, probably by ancient Christians who also carved a cross onto the pagan's forehead for good measure. Aesthetically, noseless statues are less pleasing than fully nosed ones, so in the last two centuries or so, overzealous museum curators and collectors started crafting fake stone noses to put back on their statues. Modern art-lovers don't like this fakery and crave authenticity, so a new trend of disembodying fake noses from real classical statues has now begun. At the famous Ny Carlsberg Glyptotek art museum in Copenhagen there's a rather macabre display of newly disembodied stone noses (and various other appendages) from their collection of Greek and Roman sculptures. One wonders whether there's a stray Cleopatra proboscis lurking somewhere in the collection.

A History of the World Through . . .

TRIỆU THỊ TRINH'S BREASTS

AND HOW THE PATRIARCHY TRIED TO KEEP A GOOD WOMAN DOWN

—————— OR ——————

WHAT'S MAYBE MISSING IN HISTORY BOOKS

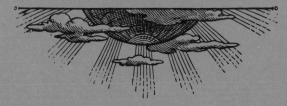

(225–248)

ADY TRIỆU (BÀ TRIỆU IN VIETNAMESE) is
notable for many things: being a woman warrior in
3rd-century Vietnam, fighting against the Chinese,
becoming the Vietnamese Joan of Arc. She's a national
heroine to this day. But when it comes to a collection of body parts
in history, she's most notable for her breasts. More specifically, her
3-foot-long breasts.

The Wu dynasty of China had ruled Vietnam since the 1st century
and, over the years, had been particularly hard on the local Shih dynasty
rulers—meaning they wanted to get them out. In 226 CE, when the
Wu Chinese made yet another concerted effort to do so, killing more
than 10,000 Vietnamese in a purge, Triệu, an orphan living with her
brother, decided it was time to fight back. The tales say that when her
brother argued that she should just get married instead, as any proper
woman would, she refused, saying: "I want to drive out the enemy
and free our people. Why should I imitate others, bow my head, stoop
over and be a slave? I will not resign myself to our usual women's lot,
bowing our heads to become concubines."

The untraditional Triệu then became a rebel warrior, leading a
thousand troops, both men and women, in thirty battles against the

Wu until she was finally defeated, and either died by suicide (some say she allowed her elephant to trample her) or just disappeared into the clouds. Or so they say.

"They"—in this case, Vietnamese folklore—also say a lot of other things, some of which sound a little implausible, some of which don't: that Chinese soldiers said they would rather face a tiger than "the Woman General in Yellow Robes," that she was 9 feet tall and could walk 500 leagues a day, that her voice was "as loud and clear as a temple bell," and that she charged onto the battlefield on an elephant, topless, her breasts knotted across her chest or slung over her shoulders.

Many scholars theorize that her breasts, the 9-foot height, and all the other amazing feats were a way for males-are-superior Confucianism to handle a nontraditional Vietnamese woman. Prior to the Chinese and their introduction of their Confucian system, Vietnamese women were treated more as equals. With the onset of Chinese rule, they slipped to inferior status. How, then, to handle a warrior woman? Make her a superwoman, an immortal, a god rather than a person, so you can stay within Confucian societal norms and have your female superhero too.

So it's not easy to determine what, then, is true about Triệu. There's a lack of historical data other than the folktales and legends passed down orally through the centuries. She is overlooked in Chinese histories (logically; to them she was just a pesky, insignificant rebel) and is included in only two early Vietnamese histories, both written centuries after she was alive. The 15th-century *Complete Annals of Đại Việt* (the official history of the Lê dynasty) mentions "a woman from the Cửu Chân commandery named Triệu Ẩu," another of the names she's known by, and tosses in that "Ẩu has breasts three thước [a thước is a little over a foot] long, ties them behind her back, often rides elephants to fight." And the *Giao Chỉ*, a.k.a. the *Imperially Ordered Annotated Text Completely Reflecting the History of Viet*, written in the 19th century, mentions a woman "with the surname Triệu, with breasts three thước long, unmarried, assembled people and robbed the commanderies, usually wearing yellow tunics, feet wearing shoes with curved fronts,

| An Ignominious Defeat

ONE OF THE MORE FAMOUS STORIES ABOUT TRIỆU HAS IT THAT, for all of her ferocity on the battlefield and her penchant for not following the path of the traditional woman, she had one traditional rap against her: She was "pure-minded." So her archnemesis General Yu Lin (a.k.a. Luc Dận) supposedly devised a rather unique if unsophisticated battle strategy that effectively got to the bottom of things: He had his troops forgo their pants in the next battle. They charged onto the field raising as much dirt and dust as possible, bottomless, genitals exposed for all to see. (Presumably this wasn't pleasant, but war, as they say, is hell.) The sight of all the male genitalia was too much for the battle-hardened yet demure Triệu. She pulled her elephant around and fled from the battlefield, and her troops were routed. Depending upon the particular story, Triệu, still unwilling to surrender, either killed herself or simply vanished into the mountains.

and fights while sitting on an elephant's head, becoming an immortal after she dies."

Yes, both "official" histories felt the need to mention the 3-thước-long breasts, albeit, well, flatly. Granted, 3-foot-long breasts slung over any woman's shoulders, even one who wasn't on an elephant, are difficult to overlook. But to fit them into the prevalent chest sensibilities of both Triệu's time and the time when the histories were written, these breasts are compelling not just because of their size, but also—and more importantly—because they are indisputably, undoubtedly, very openly *there*. And for a long time in Vietnam, most breasts weren't.

China, with its history of foot-binding (see page 201), was equally pro breast-binding, and the Chinese influence on Vietnam was unmistakable. Since the Chinese ideal of beauty was pro small breasts, so then was the Vietnamese. Large breasts were associated with peasants, shrews, women who weren't virtuous. A woman of "good breeding" who was unlucky enough to be naturally buxom would hide it by politely and modestly and quite tightly binding her breasts of her own volition and/or the volition of her husband or father, who had the final say in things. (In fact, often a woman's husband was the one who would do the actual binding since it wasn't an easy task.)

But binding wasn't just for large breasts. Small and medium ones were bound too, since unbound breasts were indecent, definitely not something for a demure upper-class woman to flaunt. Breasts were put in their place, literally and figuratively, where they could then be discreetly ignored. Even the word for breast (*vú*) wasn't spoken in polite conversation. (Cross-cultural note: This is not unlike Victorian England, where you couldn't even use *breast* when referring to the part of a chicken.)

Enter Triệu and her unmistakably unbound, unmistakably in-your-face breasts. The fact that they were so blatant could mean a number of things. On one hand, they could symbolize an overt dismissal of proper women's behavior, a pushback against patriarchal society. They could also be a statement against the class system, since a woman with large, untrammeled breasts wouldn't be from the upper classes, but a commoner—an unusual background for a popular heroine. As such, they are revolutionary breasts indeed.

But some scholars think the breasts aren't as much revolutionary as *counter*revolutionary, and were grafted onto Triệu in the later centuries to diminish her, to make her an abhorrent beast of a woman. It was one way of trying to put a woman, even a worshipped warrior woman, in her place. And then, of course, there's the thought that the breasts, along with her other attributes, were added to make her less human and more of a superhuman and so less of a threat to the established Confucian patriarchy.

That's a lot of heavy lifting for a pair of breasts, even if they're 3 feet long.

Penile Protection Against Long-Breasted Triệu

LEGEND HAS IT THAT TRIỆU CONTINUED FIGHTING AFTER HER death, just on a different plane, sending a plague to the Chinese troops and haunting Yu Lin in his dreams, so unnerving him that he needed yet another plan to repel her. The resourceful general then had hundreds of images carved out of wood to be hung over all the doors in his camp. The carved images? All penises. But he wasn't the only person who thought carved penises were imbued with powers of supernatural salvation. The concept of helpful bad luck–repelling penises had a long history prior to him, and were especially big in ancient Rome. This stemmed from the ancient Roman god Fascinus, who protected humans from things supernatural—from evil spells to the evil eye. Fascinus didn't look like the other gods in the pantheon. He didn't have a human form but rather was simply a penis. Unlike a mere mortal one, though, Fascinus the divine phallus was a penis with wings. Ancient Romans carved winged penises, called fascina after the god, to use as protective amulets or to put in a house or hang in a window. (Fascinating, if tangential, fact: The English word *fascinate* is derived from the Latin *fascinus* or *fascinum*, meaning a penis—or an evil spell.)

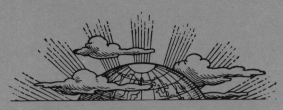

A History of the World Through...

St. Cuthbert's Fingernails

AND

HOW THE CATHOLIC CHURCH USED PIECES
OF SAINTS TO SPREAD THEIR INFLUENCE

—————————— AND ——————————

MAKE AN ARM AND A LEG IN THE PROCESS

(634–687)

 Y ALL ACCOUNTS, ST. CUTHBERT, a 17th-century monk, wasn't the type of person who would demand a regular manicure. But that's just what he did . . . several centuries after his death.

Alive, Cuthbert was a monk, a prior, and a bishop known for his good works before opting for the quiet life of a hermit until his death. Dead, he was the most popular saint of Northern England, called the "wonder-worker of England" because of the miracles that happened to those who prayed to him, particularly by his sarcophagus. His remains were equally miraculous, as 8th-century historian Bede noted. When Cuthbert's coffin was opened 11 years after his death in 687, his body was "uncorrupted," as if he had died just minutes before. Yes, like so many other saints, even though Cuthbert was dead, he wasn't *quite* dead the way a non-saint would be.

As such, even though laid to rest in the chapel in Durham, his fingernails supposedly kept growing, as well as his hair. According to contemporary accounts, an 11th-century sacristan at Durham, Alfred Westou, served as Cuthbert's manicurist and barber, keeping his nails clipped and his hair trimmed. It made the saint happy, always a good thing, since his tomb was a very popular stop for religious pilgrims, and a happy saint would be more likely to intercede on behalf of them. But there was another major upside: The fingernails and hair clippings could be sold or given to other churches or people to be used as relics. Relics were big business in the Catholic Church. Alfred himself was prone to giving out small snippets of St. Cuthbert to get small snippets from other saints as well as other relics in return.

There was a great demand for relics in the 11th century, a demand that had begun centuries before. Near the end of the 8th century, the Second Council of Nicaea, a meeting of Christian bishops, issued a decree: Every consecrated altar had to contain a relic or relics, a fragment of a saint, a piece of something connected with Jesus Christ's life, or the like. This was just the beginning of relic mania. Over time, relics became even more popular and necessary. Relics weren't only in churches and shrines anymore. They were more widespread, used to bring blessings further afield and to spread the Christian religion. Even the swords used when knights took their oaths had to have a bit of saint in their pommels. Relics could protect you, generate a miracle or two.

Initially, the Church didn't like the idea of cutting apart or otherwise removing parts of a holy corpse, but there was a technical caveat: Things that could *naturally* be detached were okay for distribution. So things like hair, teeth, and nail clippings à la Cuthbert started cropping up, suitably displayed, of course, in churches, cathedrals, and shrines for visitors to venerate. Over time, the Church got more lenient about what could be done with a holy corpse, probably because the more relics were out there, the better in terms of spreading religion and, of course, getting money.

At this point, relics had spread from the Church to royalty. Then regular people—more specifically, regular people with the money to buy them—also began acquiring relics for their personal use. In some cases, it was for salvation; in others, a way for their own home chapel or town to attract pilgrims, and money.

Even though, much later, the Council of Trent of 1563 stressed that where relics were concerned "every superstition shall be removed and all filthy lucre abolished," not many people from bishops on down really paid much attention. Relics had become big business, so the more the merrier—and the more the wealthier. Since visiting a holy shrine was the medieval religious equivalent of a trip to Disneyland, the filthy lucre was happily spread around. Traveling pilgrims spent money on lodging, food, and drink, not to mention mementos and souvenirs. Competition got quite intense as the different communities wanted their saint (if they were lucky enough to have had one buried there) or at least their bit of saint (if they had gotten a relic) to draw more tourists—and more money—than the saint or bit of saint at a community nearby. It got to the point where different places competed not only for pilgrims but also for relics. Relics would disappear from one church and pop up in another. And it could get worse. So eager were people to get their mitts on a piece of a holy person that guards sometimes had to protect a holy corpse to prevent people from doing a little sub-rosa dismemberment. (It's said that sometimes they even had to protect older holy people from murder.)

There was also the problem of relic identification or mistaken identity. Without DNA testing available, it wasn't easy to be definitely sure that one particular bone or body part actually belonged to the saint in question. After all, one clavicle or jawbone or tibia looks a lot like another. Who's to say that this particular one actually belonged to that particular person? This is how John Chrysostom, theologian and archbishop of Constantinople, wound up with not one but four "official" skulls—one in Greece, one in Russia, and two in Italy. In 1204, crusaders took his skull and other relics from Constantinople to Rome. But somewhere along the line, the actual skull got misplaced or multiplied, as the case may be, with different churches claiming they had the true relic.

Over time, the relic craze began dying out—possibly victim of its own success. It had been satirized by writers like Boccaccio and Chaucer in the 14th century, and by the 16th century, there were so many relics and "relics" out there that religious reformer John Calvin complained that there were enough pieces of the so-called True Cross to build a sizable ship.

In the case of St. Cuthbert, though, there was only the one saintly uncorrupted body. Or so it's thought. When the English Reformation swept across the country, the cathedral at Durham was plundered and St. Cuthbert's tomb opened. But he wasn't in it any longer; apparently the monks had hidden it. St. Cuthbert remained lost until 1827, when a secret tomb was discovered in the cathedral holding, not the perfect life-in-death body from before, but the more typical non-saintly decayed corpse. No one can be sure who it is, but it is marked as St. Cuthbert in Durham Cathedral, where pilgrims and worshippers can still visit the Wonder Worker.

While not as much of a religious focal point as in the Middle Ages, relics are still displayed for pilgrims and worshippers in numerous shrines and churches. And they're still available for purchase, but with a distinct 21st-century spin. People no longer have to go on pilgrimages to track down relics. A Google search for "religious relics for sale" returned a result entitled "Mary Magdalene Relic on eBay—Seriously, We Have Everything." And apparently they do. There are dozens of religious relics on the site, everything from a piece of Mary Magdalene or a bone chip of St. Alphonsus Liguori to a grab bag of 36 first-class relics of martyrs. (The listings all helpfully note that they are pre-owned, lest you mistakenly think you're getting an unused one.)

DIY Relic Acquisitions

SOME PEOPLE WENT TO GREAT LENGTHS TO GET THEIR HANDS (or other body parts) on a relic. One of the most famous relic entrepreneurs was Portugal's Doña Isabel de Carom. In 1554, she went to see St. Francis Xavier's body in Goa, where it was displayed in the cathedral. She bent to piously kiss his foot . . . and instead bit off the little toe on his right foot. She then absconded with the toe to Portugal, where she enshrined it in her own family chapel, a way to cash in on the lucrative pilgrim business. The toe remained a bone of contention, until many years later when Goa became part of India, and the toe was finally given back to be with the rest of Francis Xavier's body.

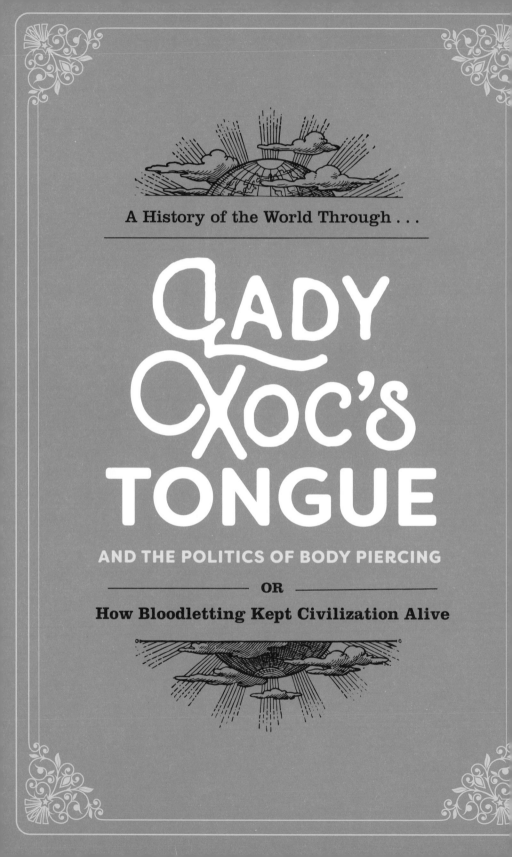

A History of the World Through . . .

CLADY CXOC'S TONGUE

AND THE POLITICS OF BODY PIERCING

—————————— OR ——————————

How Bloodletting Kept Civilization Alive

(FL. 700)

I N A DARK ROOM DEEP in the recesses of a temple in
the jungles of Chiapas, Mexico, more than 1,300 years ago,
Lady K'abal Xoc (pronounced "shoke") of the Mayan city
of Yaxchilán was performing a very bloody and extremely
painful ceremony.

As her husband, the "Blood Lord" and king of Yaxchilán, held a
burning torch and watched, Lady Xoc quickly pierced her tongue,
probably with a stingray spine, and then ran a knotted rope flecked
with shards of sharp volcanic glass through the hole. These Mayans
weren't kidding around—they were after maximum blood, and the
tongue, which is amply supplied by the lingual artery, is very rich in
it. Not surprisingly, blood poured from Lady Xoc's tongue, staining
her cheek ornaments, dripped down her face, and fell in splats onto
paper strips in a basket. The bloody paper would soon be burned as
an offering to the gods while Lady Xoc would go into a hallucinatory
trance and meet a vision serpent. It was all in a day's work to ensure
her city a good future.

The ritual is recorded on the limestone lintels of what is thought
to have been Lady Xoc's ceremonial home and tomb in Yaxchilán,
the city they ruled on the banks of the Usumacinta River in the state

Burning a Few Books

WHEN BISHOP DIEGO DE LANDA BROUGHT THE SPANISH Inquisition to the Mayans, he also brought along the "useful" practice of torture, although he allegedly claimed that bloodletting was absolutely "prohibo," and that whipping and hoisting were more the norm. Others begged to differ. De Landa is best known, however, for another atrocity: burning all the Mayan books he could find. Like the bloodletting ceremony of Lady Xoc, we have the exact date for de Landa's big bibliocide: July 12, 1562. In de Landa's words, "We found a large number of books in these characters [Mayan hieroglyphs] and, as they contained nothing in which were not to be seen as superstition and lies of the devil, we burned them all, which they [the Mayans] regretted to an amazing degree, and which caused them much affliction." Much affliction, we may add, not only to the Mayans back then but also to modern scholars now. We have only three unburnt Mayan books (codices). Much about this fascinating civilization will never be known.

of Chiapas. We know the details of the ceremony. We know the date that Lady Xoc cut into her tongue—5 Eb 15 Mac, or October 28, 709 CE—because it too was carved in the limestone. And we know that Lady Xoc was probably one of the most powerful and important women in Mayan civilization and her husband, Blood Lord (a.k.a. Itzamnaaj B'alam II or Lord Shield Jaguar), one of its most important kings.

Until Mayan hieroglyphs were deciphered, virtually nothing was known of Lady Xoc and Lord Shield Jaguar, and almost nothing was known of ch'ahb', the bloodletting ceremony. So speculation abounded as to what exactly was going on in the ornately carved depictions on tomb walls. In fact, after examining a number of carved portrayals of similar ceremonies, a certain not particularly astute but very prolific 1960s author somehow posited them as proof of extraterrestrials visiting Earth. The author was wrong (in fact, he seemed to see alien visitations virtually everywhere), but then again, his book was a huge bestseller. In the 1970s, though, dramatic breakthroughs came in reading the complicated Mayan hieroglyphs, and the carvings of Lady Xoc drawing blood were studied in depth.

Lady Xoc's building was a mansion of many doors and many lintels, but for the bloodletting ceremony, lintels 24, 25, and 26 are the ones that matter. (Two of them were found by British archeologist Alfred Maudslay in the late 1800s; he characteristically promptly shipped both off to Mother England; the third managed to remain in Mexico.) The first (and most graphic) lintel of the trio shows the bloody tongue-perforating ceremony itself. It wasn't always tongues that were slashed in these ceremonies. Men would perforate their penises, and both sexes also cut earlobes, nostrils, and lips. Clearly, Mayans were not aichmophobes (people afraid of sharp objects). These practices horrified the early Spanish conquistadors who had seized the Mayan lands in the 1500s.

Of course, the Spanish had their own form of (approved and holy, of course) bloodletting, which they practiced during the Spanish Inquisition and included such "civilized" practices as ripping off arms via the rack, cutting flesh with pincers, and using ever-handy thumbscrews; some say they brought these techniques to their Inquisition in the Yucatan. Rather ironically, one conquistador Bishop Diego de Landa disapprovingly described the voluntary blood ceremonies in his *An Account of the Things of Yucatan*:

> At times they sacrificed their own blood, cutting all around the ears in strips which they let remain as a sign. At other times they perforated their cheeks or the lower lip; again they made cuts in parts of the body, or pierced the tongue crossways and passed stalks through, causing extreme pain; again they cut away the superfluous part of the member, leaving the flesh in the form of [s]cars. It was this custom which led the historian of the Indies to say that they practised circumcision.

But why were they doing this? The answer is partly found in the next lintel, number 25, which shows the aftermath of the bloodletting ceremony. Lady Xoc, holding two baskets of blood-soaked papers and presumably in intense pain, dizzy from the loss of blood, and hallucinating, stares up at the part-snake part-centipede "vision snake," which is opening its jaws to reveal the materialization of the founder of the royal line. Below is Chaac, the fierce Mayan storm god. The blood ceremonies were literally linked with political legitimacy as bestowed by the heavens.

At first, that aspect doesn't seem too different from other such sacrificial narratives in other parts of the world, but there's more. Lady Xoc not only sees the gods and communicates with them, but she also literally feeds them with her blood in a kind of reciprocal repayment. And while no one is sure, it also seems that in the final lintel, number 26, Lady Xoc with her visions is now busy empowering her husband for battle. (Incidentally, any citizen or enemy doubting Lady Xoc's divinely inspired powers would have been in for a major disappointment; it seemed that Lady Xoc's bloodletting really worked to energize her husband. Hubby Blood Lord was busy fighting—and winning—many battles well into his geriatric 80s. He died in his 90s, still in charge. Blood will tell.)

But why all that blood? There's no one answer, but Meso-American life was literally steeped in blood. The Aztecs who ruled further north are the most famous—ripping out the hearts of sacrificial victims tends to get a lot of attention—but the Mayans weren't averse to this type of bloody sacrifice either. The key reason: the belief that the gods had given up some of their divine blood to bring life to humans, so, as with Lady Xoc, humans had to give back to the gods, feed them with blood, to maintain the order of the universe.

A recent study found that Mayan bloodletting ceremonies were concentrated in southern Mayan lands and seem to have been particularly prevalent at the end of the Classic Period, just when Mayan civilization was beginning to collapse. According to one scholar: "Maybe when they see their world crumbling down around them, they are frantically trying to communicate with the gods around them."

Or maybe they're just showing off to their enemies. After all, you had to be pretty tough to rip through your tongue with volcanic glass shards.

And Now
for the Aztecs . . .

YES, YOU MIGHT SAY THAT MOST OF THE GREAT CIVILIZATIONS of Middle America—the early Olmecs, the Toltecs, the Aztecs, and the Mayans—had a thing for blood. (And so did the West: but Roman, Greek, and Near Eastern sacrifices tended to throw in charred meat as well.) The Aztecs are probably the best known today. But let's get one thing straight: The Aztecs didn't just confine themselves to making blood-soaked sacrifices on pyramid tops as modern films and novels might have it. Like the Mayans, they also practiced personal ritual bloodletting, although unlike the Mayans they probably used bone or maguey spines as well as obsidian. But like them they burned the blood-soaked paper strips as offerings.

On the sacrifice front of things, the Aztecs offered up animals as sacrifices but are best known for their human sacrifices. These surprisingly often involved a sense of restraint. A type of warfare called xochiyaoyotl, or flowery war, was usually conducted to obtain captives for sacrifice; once the requisite number were captured, the war was ended. No Hundred Years' War à la Europe in this civilization.

The bravest and/or handsomest captives were then selected for the rather dubious honor of being taken to the top of a pyramid, stretched out over a special stone platform, and having their chests cut open with an obsidian or flint knife and the bloody heart removed. Although at this point it presumably didn't matter to the victim, his or her heart would then be placed in a special stone vessel (a cuauhxicalli) or a chacmool (a reclining stone figure with a bowl at the waist) and burnt in offering to a particular god. Alternatively, a victim could be decapitated, dismembered, skinned, made to play in a ball game (losers all killed), burnt (heart then removed), or killed in the Aztec equivalent of a Roman gladiatorial contest with rather unfair odds—the victim was tied and given a feathered club; opponents were given sharp obsidian swords—which brings up an interesting parallel. The Roman gladiatorial games apparently had a religious origin like that of the Aztecs; the Roman precursors the Etruscans originated these combats as a religious funeral rite.

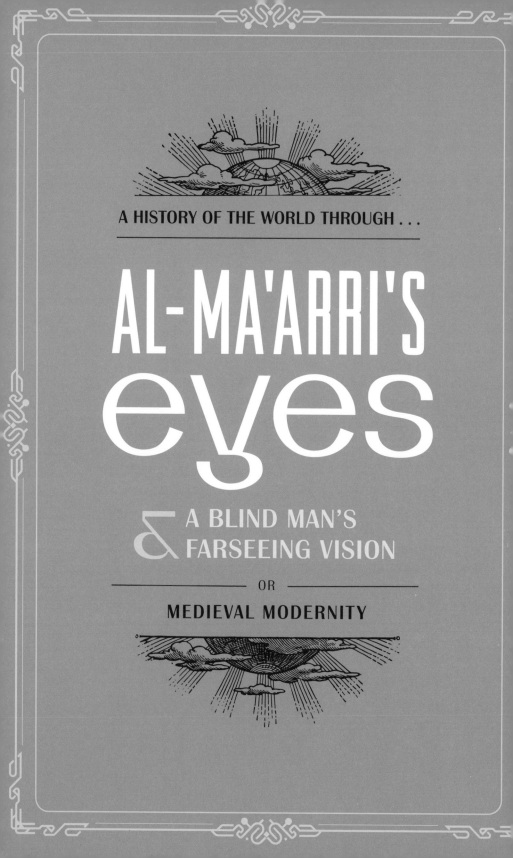

A HISTORY OF THE WORLD THROUGH . . .

AL-MA'ARRI'S
eyes

& A BLIND MAN'S
FARSEEING VISION

—— OR ——

MEDIEVAL MODERNITY

(973–1057)

N THE SPRING OF 2013, members of the local Syrian branch of al-Qaeda fulfilled their yearlong dream: They beheaded a statue.

The decapitation was hardly an art crime. The bust, according to many art historians, bordered on the kitsch—a blind man in a turban with piercing-looking eyes that didn't look at all blind. It was a highly idealized depiction (think statues of generals in the United States), sculpted in 1944 for a celebration of the blind man's work. He was poet Abū al-AlāʾAḥmad ibn Abd Allāh ibn Sulaymān al-Tanūkhī al-Maʿarri, better known as al-Maʿarri. (This part of his name refers to his hometown near Aleppo in Syria, which is where the bust had been placed to honor his work and where the itinerant al-Qaeda members did their malicious metalwork.) He was famous and infamous in Arab literary and philosophical history. And he had been dead for more than 1,000 years.

So why chop off the metallic head of a long-dead poet? For one thing—and for many, such as the al-Qaeda decapitators, the main thing—was that besides being a great poet, al-Maʿarri had been called blind to religion, an Islamic heretic, even an atheist. This view, though, has been disputed by some scholars, particularly by Oxford professor

D. S. Margoliouth. What is indisputable, though, is that he was one of the greatest of Arabic wordsmiths, a master of Arabic poetry. And he credited his eyes for his success.

The defining moment in al-Ma'arri's life came early, when he was four years old and had a devastating attack of smallpox that blinded him. According to a contemporary who visited al-Ma'arri: "The disease that attacked him in his boyhood had left its deep traces on his emaciated face . . . I looked into his eyes and remarked how the one was horribly protruding, and the other, buried in its socket, could barely be seen."

A tragedy, yes, but al-Ma'arri later said that his blindness gave him an ear for words and a talent for memory. This idea of one sense compensating for another's loss was a big thing in the Middle Ages. Probably because so many people lost limbs, eyes, or other portions of their anatomy from disease, warfare, or accident, their concepts of disability differed from our often more pejorative take today. By all accounts, al-Ma'arri's memory was prodigious. One apocryphal example has al-Ma'arri perfectly recalling a conversation held in the Azerbaijani (other stories say Persian) language, which he didn't even know. Another story has him being told to duck on a path while riding to avoid a low-hanging branch; two years later on the same road at the exact same spot, he automatically ducked. (There was no need; the tree had been cut down.) Obviously, a memory like this put him in good stead while writing.

His lack of sight didn't affect his worldview either. He was intellectually farseeing, a philosophical poet who anticipated (and actually probably influenced) the medieval West's greatest poet, Dante. Al-Ma'arri's *Risalat ul Ghufran,* or *The Epistle of Forgiveness*, is a divine comedy on Arabic civilization, focusing on poetry, and yes, some say that this "divine comedy" inspired Dante's *Divine Comedy*. His *Spark of Flint* (or *The Tinder Spark*), the *Sakt al-Zand,* is his most famous work and won him accolades as a master of Arabic classical poetry. Finally, his *Luzumiyat* shows his extreme poetical prowess—every verse in the

Arab Ophthalmologists

THE MIDDLE EAST WITH ITS BRIGHT SUNLIGHT AND SANDY WINDS was not the best place for optimal eye health. The Arabs became pioneers in ophthalmology, including (wince) the use of an injection syringe, invented by Ammar ibn Ali of Mosul, which when jammed into the cornea of the eyeball could extract soft cataracts. But, in al-Ma'arri's case, along with other sufferers of ocular smallpox complications, it wouldn't have helped. The smallpox virus often causes extensive ulceration of the cornea, causing its complete destruction. Modern antiviral and steroid applications applied early in the course of the disease could have helped, but these treatments were 1,000 years in the future.

poem rhymes in two disparate consonants instead of one. Here too his famous ear outshone his sightless eyes.

His poems attracted a great deal of attention not only for their beauty, but also for their unorthodox content. In one poem, al-Ma'arri boasts that his verses could rival those of the Quran. In addition, many of the more pious statements in this (and other works) are exaggerated, so much so that they might be considered heresies cleverly disguised as piety. These sorts of things didn't exactly endear him to the more dogmatic religious types of whom there were many, even in the enlightened Islamic world of the day. But al-Ma'arri wasn't fazed. When anyone accused him of being a heretic, he simply said it was envy speaking. Not only did al-Ma'arri never get formally charged with heresy, but an Islamic judge even said that al-Ma'arri recited the Quran in such a way that no one could disbelieve his faith. A small industry defending al-Ma'arri's faith ensued, which has continued into modern times.

To many, al-Ma'arri seems to be the prototypical deist along the lines of Thomas Jefferson or Benjamin Franklin: denying or downplaying divine revelation, seeing God more as an impersonal creator rather than a deity getting involved in the nitty-gritty of life, and railing firmly against any religious sectarianism or claims to exclusive truths. So along with al-Qaeda, al-Ma'arri would have probably gotten under the skin of Jerry Falwell or, for that matter, all religious fundamentalists from all other religions. We can safely assume this would not have bothered him at all.

That a controversial poet-philosopher like al-Ma'arri not only lived but also thrived in his medieval era (he lived between 973 and 1057 CE) says a lot about the state of the medieval Islamic East. It welcomed, or at least tolerated, intellectual discussion, differing interpretations, and civilized argument, and it was a civilization in which a blind person wasn't handicapped by a lack of sight. In al-Ma'arri's time, the greatest cities were Eastern: Baghdad and Cairo were among the largest cities in the world. Shared language and legal systems lowered transaction costs and stimulated commerce, and transport via established caravan routes among the cities was (relatively) swift and easy. As such, peripatetic scholars could visit libraries across the Islamic world and seek out competing intellectual points of view. Meanwhile, mostly quasi-literate and localized Western Europeans were focusing on more prosaic things like how to fatten a scrawny pig for the winter and whether and when they should take their yearly bath. Of course, not too long after al-Ma'arri's death, it was the West's turn to opt for urban revolution and intellectual dynamism—in many cases building upon ideas from the Islamic East, which in turn had built on ideas from the Greeks and Romans, who in their turn had built upon ancient Egypt and Babylon.

Al-Ma'arri, the Proto-Vegan

LONG BEFORE OAT MILK AND SOY CHORIZO WERE AVAILABLE IN your local supermarket, al-Ma'arri became a vegan. It was a philosophical decision. He had come to believe in the sanctity of all life, so decided to eschew not only meat but all animal products, including honey from bees, like any good modern vegan. And like any good poet, he put his vegan manifesto into verse:

Do not unjustly eat fish the water has given up,

And do not desire as food the flesh of slaughtered animals,

Or the white milk of mothers who intended its pure draught

For their young, not noble ladies.

And do not grieve the unsuspecting birds by taking eggs;

For injustice is the worst of crimes.

And spare the honey which the bees get industriously

From the flowers of fragrant plants;

For they did not store it that it might belong to others,

Nor did they gather it for bounty and gifts.

I washed my hands of all this; and wish that I

Perceived my way before my hair went gray.

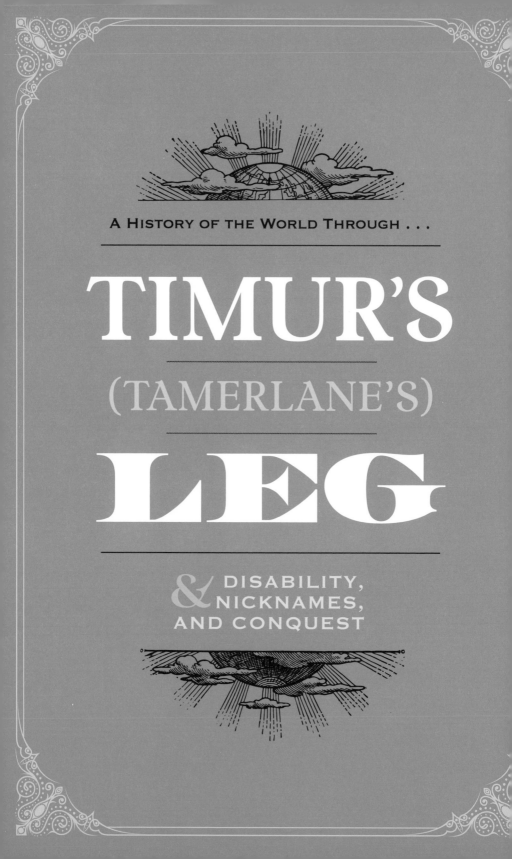

A HISTORY OF THE WORLD THROUGH . . .

TIMUR'S

(TAMERLANE'S)

LEG

& DISABILITY,
NICKNAMES,
AND CONQUEST

(1336–1405)

OMETIME IN THE EARLY 1400s, a famous Uzbek
poet named Sakkaki wrote an odd poem about a young
man intently watching a "crippled ant" trying to walk
despite its obvious ambulatory deficiencies. The young
man in the poem just so happened to have been badly injured in his
right leg, a serious problem for someone born in a warrior culture.
He was so inspired by the plucky little ant that he too decided to per-
severe despite this disability. And so he did. The man in the poem is
Timur, one of the world's greatest—and most notorious—conquerors.

It's a cute little tale about a man and his destiny, even if it's probably
untrue. (There's a similar apocryphal story about another king in far-off
Scotland, Robert the Bruce, who was encouraged by the persistence
of an even leggier spider.) Timur is surrounded by vagueness and
contradiction when it comes to stories about his life.

But we do know the basics of his conquests: He succeeded in conquering most of the Islamic world and was striving to take China when
he died. In death, his huge empire collapsed and fragmented, but
his legacy remained. Among other things, his descendants created a
high intellectual and artistic culture. (One built the Taj Mahal.) But
Timur's legacy has a dark side. This great Islamic conqueror, arguably

as "great" as Alexander the Great and Genghis Khan, had a reputation for incredible cruelty and fearsomeness—so much so that his reach was said to extend beyond the grave.

The fear and hatred that Timur aroused was justified for his enemies and even among his own people. (A European ambassador visiting him noted that one of his servants who had served dinner late just missed having the typical punishment of having his nose pierced "like a pig.") In building his great empire, Timur slaughtered millions—some say upwards of 17 million—often in horrifying and inventive ways, including being buried alive, cemented into walls, sliced in two at the waist, and trampled to death by horses, as well as your more standard modes of beheading and hanging. As he crossed central Asia, he left pyramids of skulls in his conquering wake.

Timur's Persian enemies not surprisingly loathed him so much that they homed in on the one easy thing they could mock him for (presumably at a safe distance): his not-so-perfect right leg. The Persians added *lame* to his nomenclature, as in Timur-i-lang: Timur-the-Lame, or Tamerlane. Others called him "the Iron Cripple." Both of those pejorative nicknames entered history.

By most accounts, Tamerlane walked with a very distinctive limp. But here's where some vagueness or contradiction comes in: Tamerlane was a master of public relations; he strove to create an image of superhuman ability, of cruelty mixed with wisdom (he was by all accounts an intelligent and cultured man, although illiterate), of profligacy mixed with temperance. Most important, he projected the image of a man who had risen out of humble beginnings to become a mystically endowed leader who could see and do far more than mere mortals. And so the question arises: Did Timur use his disabled right leg as just another part of his image-building campaign—i.e., poor disabled boy makes good?

We'll never know for sure, of course. We do know that Timur was an intelligent, crafty, and ruthless tactician who was well aware how images create realities. His cruelties to conquered populations (usually reserved for those who resisted rather than surrendered) can be

The Curse of Tamerlane

TAMERLANE'S MODERN UZBEK DESCENDANTS, UNHAPPY ABOUT their beloved ancestor's exhumation by the godless Soviet communists, apparently started a whispering campaign about a legendary dreaded curse. The great warrior had supposedly written that anyone violating his tomb would be attacked within three days; others say this was written: "Whomsoever opens my tomb shall unleash an invader more terrible than I," without stipulating the exact timing of revenge.

And wonder of wonders, just after Soviet scientists opened the tomb, the Soviet Union was attacked by Nazi Germany. Was the curse true? As if to further the point, once Tamerlane's body had been belatedly returned to the grave a few years later, the Soviets had their first major victory in the war when they surrounded and defeated the Germans at Stalingrad. It's an interesting story about a graveyard curse—and it's been retold numerous times, even in the West (check out the film *Day Watch*)—but the reality is a lot more prosaic. The Nazis attacked within two days, not three as per the curse (although that's quibbling), but even allowing for premature curse fulfillment, the real problem was that, according to the latest scholarship, no real written curse on the tomb was ever actually found, just some Arabic religious texts.

seen as calculated tactics to encourage the others to submit. But he could be cruelly capricious. Once, in Sivas in central Turkey, Timur promised no bloodshed if the inhabitants surrendered; when they did, he promptly buried 3,000 prisoners alive. As he pointed out, there was no bloodshed, just suffocation. But even here, we can't be sure what is actually true and what is not. Many histories about him were written by his enemies and so painted him in the worst possible ways. Even those who supported him wanted to convey the image of a ruthless, inevitable winner, showing that he was truly the baddest ass in town—tougher even than the terrifying Genghis Khan.

Timur embraced Genghis Khan comparisons, and took them a step further, claiming he was a descendent of the great leader. As Genghis's heir, naturally other central Asians should rally around him; all he was doing (supposedly) was restoring Genghis's rightful rule. It was a nice PR move that helped cement his legitimacy, but was almost certainly fiction. In reality, Timur was born to minor nobility unrelated to the line of Genghis in central Asia, in what is today Uzbekistan. Early on, he showed his stuff as a proto-conqueror, taking control of a local kingdom, or khanate. There was no stopping him from that point on: His ever-expanding multicultural military conquered the Turkish-speaking areas surrounding his homeland: Syria, Turkey, and India.

In another shrewd move, he was careful not to usurp traditional titles once he was in control of the new territories. He took the more modest official title of commander (amir, from which, incidentally, the English word *admiral* derives) but also the flashier "Sword of Islam" and Sahib Qiran (Lord of the Fortunate Conjunction) to show who was really in charge and to play up divine inspiration.

Even the origin of Timur's leg injury is the subject of debate. Some sources say it was the result of being shot at after a youthful bout of sheep rustling; others say it was a more respectable battle injury suffered when fighting for the khan of Sistan. Maybe Timur liked both stories to play up different sides of his nature, which leads us to the question: For all the discussion of his injury, just how lame was he?

The answer came in 1941, when the Soviet Russians, whose ances-
tors had conquered much of central Asia, decided to open the Guri
Amir mausoleum in Samarkand where he was said to be buried. Timur's
body was in a crypt under the mausoleum, beneath a huge slab of
nephrite—dark green jade—supposedly the largest jade piece in the
world. (It had been broken in half by the later Persian ruler Nadir
Shah who had hoped to bring it back home.) The Soviet anthropologist
Mikhail Mikhaylovich Gerasimov was in charge of the exhumation,
and he found what clearly looked to be Timur's remains. His team
of scientists proceeded to clean silt and saline crystals from Timur's
skull, remove bits of hair (reddish brown) as well as skin and brain,
and put them into bags for the anatomy department at the Samarkand
State Medical Institute. The result was a scientific dismantling of the
body . . . but not of the legend. The cadaver showed evidence of injury;
pathology reports indicated "tuberculous cavities in the right femur
and tibia with bony fusion of the femur and patella, and complete
bony ankylosis of the right humerus and ulna."

In other words, Timur had injuries due to wounds in the right leg,
probably had a stiff right arm, and appeared to have lost two fingers.
He probably walked in a stiff-armed, slightly hunched, shuffling gait.
The Persians were right.

How Timur Spawned the New Science of Forensic Facial Reconstruction

WHEN SOVIET ANTHROPOLOGIST MIKHAIL M. GERASIMOV OPENED Timur's tomb, he wanted "to gaze on the faces of those long dead." This became his life work—he was a pioneer of scientific forensic sculpture, the recreation of faces. To make the long-dead face come alive, Gerasimov would take a skull and slather a mixture of beeswax, modeling clay, and colophonium (rosin) on it, sculpting a realistic-looking face that (hopefully) actually looked like the person. It wasn't easy. Not only did he need a highly detailed knowledge of facial muscles, but he also had to deal with the technical difficulties of reconstructing extensive soft tissue areas such as noses and eye sockets.

Timur was his big break. From there, he went on to reconstruct the faces of more than 200 people, including Ivan the Terrible. His skull-examination techniques were also used to identify the remains of Tsar Nicholas II's family. Followers of his techniques have since reconstructed the face of Tutankhamun, and, most recently, of Jesus Christ.

A History of the World Through . . .

RICHARD III's BACK

&

HOW A HISTORICAL PUBLIC-RELATIONS
IMAGE CAMPAIGN—LATER AIDED BY
A CERTAIN BIG-NAME PLAYWRIGHT—
CREATED A VILLAIN

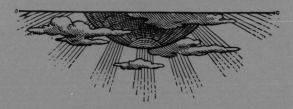

NGLAND'S KING RICHARD III is known for being the last king of the Plantagenet dynasty. But there's a specific reason he appears in a collection concerning body parts.

Yes, England's King Richard III was said to have kyphosis, or, to use the problematic term, a hunchback. In fact, to the public at large, he is probably the world's second most famous person with this condition, a step behind the fictional Quasimodo, hero of Victor Hugo's 1831 novel *The Hunchback of Notre Dame*. And just as Quasimodo got even more famous as "hunchback-with-a-heart-of-gold" after he got Disneyfied, Richard III got even more infamous as "hunchback-with-a-heart-of-darkness" after he got the Elizabethan version of the Disney treatment: the Shakespeare treatment.

First, the nonfiction: Richard III's life was inextricably tied to the Wars of the Roses, the series of battles to win and keep the British throne. They were fought between the two Plantagenet branches, the Yorks (initially led by Richard's father, the Duke of York) and the Lancasters (initially led by King Henry VI). Ultimately, Richard won the throne for the Yorks, but through a rather sideways manner—through taking (some say usurping) his 12-year-old nephew Edward's throne while lord protector. And here the history gets murky. Some say Richard worked behind the scenes to get Parliament to say his nephew Edward (and his younger brother) were bastards, so ineligible for the throne. Many even say Richard had them killed. Others say that's all nonsense. No one is sure, but what we do know is that some people were not at all happy about Richard being king, and they found a willing champion to fight Richard in Henry Tudor (later Henry VII). The battles for the throne started again. And Richard, who had held the throne for only two years, was killed in the Battle of Bosworth Field, marking an end to the Wars of the Roses and the beginning of the Tudor dynasty, the end of the Plantagenets and the creation of a quasi-fictional "hunchbacked" villain.

Roughly 100 years after Richard's death, in about 1591, Shakespeare got his mitts, or rather pen, on him. Richard appeared in William Shakespeare's *Henry VI, Part 2* (in which he is dubbed "Richard Crook-back") and *Henry VI, Part 3* (in which he soliloquizes about his "disproportionate" appearance, with an arm "like a wither'd shrub," an "envious mountain" on his back, and legs "of an unequal size"). The eponymous play *Richard III*, written a couple of years later, solidified his fame as a villainous hunchback, an identity that has persisted into the 21st century.

Surprisingly, the word *hunchback* didn't even appear in *Richard III*. Then again, it really didn't have to in order to get the point across. It loomed large throughout the play. Richard is called a "lump of foul deformity," an "elvish-mark'd, abortive, rooting hog!" "that bottled spider," and the twin toad insults: a "foul bunch-back'd toad!" and a "poisonous bunchback'd toad."

In the play, it's made clear that Richard was a schemer, eaten up by ambition and determined to get the throne after his brother's death, to the point where he was willing to kill his nephews. There's no mention of his many attested good works or of his popularity with the lower classes. Shakespeare does paint him as a valiant warrior in the final battle that led to his death at the hands of the Tudors, but that's about as far as he goes on the positive side of things.

Shakespeare's *Other* Smear Job

SHAKESPEARE ALSO DID A MAJOR CHARACTER OVERHAUL ON Macbeth. Yes, Macbeth was king of Scotland and did kill King Duncan to ascend to the throne . . . but that's where the correlation between the facts and the play end. He wasn't just an ambitious man goaded by an even more ambitious wife; he had a legitimate claim to the throne through his mother's side. His cousin Duncan wasn't much loved and was considered quite a disaster as king. Macbeth killed Duncan in a battle; he wasn't wracked with guilt, didn't go on a killing spree, and actually ruled for 17 years, not just one, as Shakespeare had it.

So why did Shakespeare change the story of these two not-terribly-well-known Scottish kings? Just as he favored the Tudor side of stories while Elizabeth I was on the throne, with *Macbeth* he was brownnosing King James I, who just happened to be a descendent of Duncan and believed in the divine right of kings. By making Duncan's loss of the throne unfair, Shakespeare was underscoring James's right to the British throne.

From then on, there was really no going back where Richard's public image was concerned. Yet, everyone also accepted that, as Shakespeare depicted, he was quite a successful warrior until he died in the last major battle of the Wars of the Roses. It didn't quite jibe. For good reason: Richard *didn't* have a withered arm, his legs were the same length, and he didn't even have kyphosis.

This all came to light when a 500-year-old skeleton was found in a not-terribly-royal place: a parking lot in Leicester, England, which was the site of Greyfriars friary, where Richard's body had been taken and buried after he was killed in the Battle of Bosworth Field. When the parking lot was excavated in September 2012, a skeleton of a man who had apparently died of battle wounds was unearthed. Tests showed the bones were of someone roughly 30 or so, were about 500 years old, and, most tellingly, had mitochondrial DNA that matched two matrilineal descendants of Richard's sister Anne of York. A few months later, researchers announced that this was, beyond a reasonable doubt, the body of Richard III.

The takeaway: Richard wasn't the reviled "humpback" of Shakespeare's plays, but he didn't have the straight spine his modern defenders claimed he had either. The truth was somewhere in the middle. The middle of Richard's spine was curved, showing that he probably had adolescent-onset idiopathic scoliosis, which often occurs during a teenage growth spurt, but his vertebrae were normal, as were his arms and legs. So while his right shoulder was just a bit higher than his left, everything else about him was like your average Plantagenet. (And, according to the DNA, there was a 96 percent probability that he was actually blond and blue-eyed.)

So how did Richard III get to be so inaccurately depicted? Thank (or blame) the Tudors and a particularly effective negative-spin campaign. Since the Tudors were taking the throne after a long line of Yorks, they wanted to be sure the public saw them as saviors and Richard as a monster. It's one of the most famous—and most successful—examples of the old "the victors get to write the history" trope. Add to that the fact that Richard was the last York and last Plantagenet, and last rulers in royal dynasties tend to get a bad reputation anyway, even without a victor's PR campaign; poor Richard couldn't win.

The Tudors wanted to ensure their newly established dynasty came out smelling like a (non-York) rose. Historians and chroniclers of the time fell into line. It wasn't exactly sub rosa. Blatant case in point: When Richard III was king, historian John Rous praised him in his *Rous Roll*, calling him "a myghti prince" with "bounteous grace." A few years later, when Henry VII was on the throne, Rous wrote a bit differently about Richard in his *Historia Regum Angliae*, talking about Richard's unnatural birth, how he was "'retained within his mother's womb for two years and emerging with teeth and hair to his shoulders," and had "unequal shoulders, the right higher and the left lower." Others chimed in, particularly playing up his back.

This was particularly inspired since it literally built on Richard's scoliosis. It's said that after Richard was killed, his nude body was paraded through town thrown over the back of a horse, allowing people to thus see his crooked spine. It was only a half-step to inflate that into a full-fledged hump on his back. So-called hunchbacks were reviled at the time. They were thought to be people with a base character and were usually ostracized at best. Tossing the rounded back into the mix of other negative characteristics assured a negative public reaction.

Thomas More picked up the anti-Richard ball and ran with it in his *History of Richard III*, written between 1513 and 1518 (although he seems to have gotten a little confused as to which shoulder was actually higher), saying he was "little of stature, ill fetured of limmes, croke backed," and born with teeth. (It's probably mere coincidence that More later served on Henry VIII's Privy Council and became his lord chancellor.)

And then, of course, Shakespeare (writing in the time of a certain powerful Tudor queen) built on the prior "factual" accounts of Richard, particularly More's, and the die was cast, the hunch attached, and Richard became the deformed monster of legend.

Another Person Connected to the Tudors Who Got the Negative Physical Spin

RICHARD III WASN'T THE ONLY TUDOR CONNECTION WHO GOT A posthumous smear job. Anne Boleyn, wife of Tudor king Henry VIII, also received a negative makeover, courtesy of Nicholas Sanders, scholar, historian, and (most important) proud Catholic. In his book *De Origine ac Progressu schismatis Anglicani* (also known as *The Rise and Growth of the Anglican Schism*), he writes:

> Anne Boleyn was rather tall of stature, with black hair, and an oval face of a sallow complexion as if troubled with jaundice. She had a projecting tooth under the upper lip, and on her right hand six fingers. There was a large wen under her chin, and therefore to hide its ugliness she wore a high dress covering her throat.

Lest he seem overly critical, Sanders added that she was "amusing in her ways, playing well on the lute, and was a good dancer."

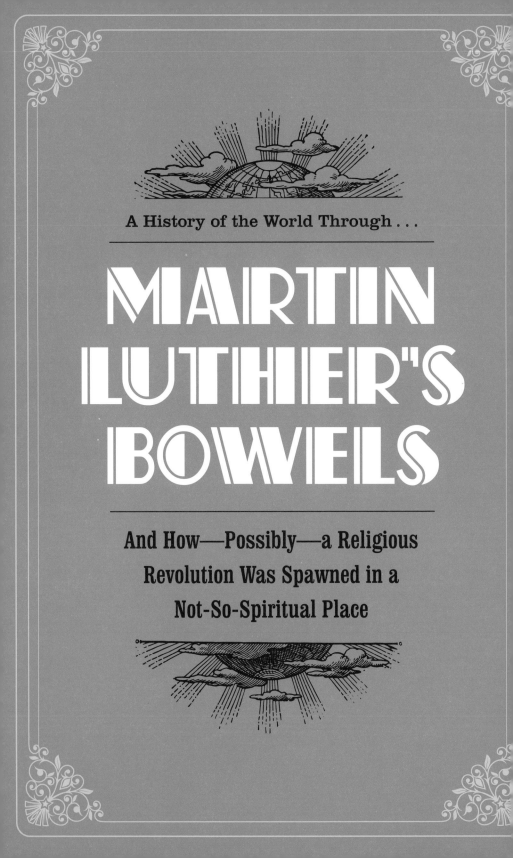

A History of the World Through . . .

MARTIN LUTHER'S BOWELS

And How—Possibly—a Religious
Revolution Was Spawned in a
Not-So-Spiritual Place

(1483–1546)

I T SEEMS SOMEWHAT SACRILEGIOUS TO imagine the prime founder of the Protestant Reformation pushing out his revolutionary new religious doctrine while straining on the toilet, but that's exactly what happened—at least according to its most important (and frequently constipated) founder, Martin Luther himself.

The Protestant Reformation of the 16th century was one of Christianity's most important and controversial movements with far-reaching repercussions. It split Western Christianity in two, spawned religious wars (including the still-simmering Northern Ireland conflict), brought on the Catholic Counter-Reformation, promulgated a new way of looking at Christian salvation, promoted individualism, reduced the power of the Catholic Church, encouraged capitalism—the list goes on and on.

And it all began . . . on a toilet?

Well, controversially. At least in part. Martin Luther was an idealistic, tormented young monk when his part in the Reformation began in Wittenberg, Germany. Young monk Luther was particularly horrified by the burgeoning practice of the Church selling indulgences. The idea was simple and theologically profitable: Sinners could literally buy their way out of after-death suffering in purgatory, where sins were burned away prior to heavenly ascent. All you had to do was pay a church representative a specified sum; he (always a man) would hand over a printed form (an indulgence) with your name filled in and the number of years off of purgatory you had paid for. To make matters worse for sinners—and what annoyed Luther even more—was that in 1515 the Church canceled the powers of their old indulgences for 8 years and declared sinners had to buy new ones. No need for confession or contrition, though, just pay up and you'd be off the hook. Pay enough and you'd be let off for even the worst of sins. Johann Tetzel, a Dominican friar, who enthusiastically peddled these time-off indulgences in Luther's Wittenberg neighborhood, allegedly boasted that for the right price he could get someone time off even if they'd raped the Virgin Mary.

Pretty terrible stuff as far as spirituality goes. Luther contemplated this and other churchly incongruities during his long-tormented hours on what the Germans called das Klo. Luther was indeed a late-medieval poster boy for constipation. His letters complain of cramps, a sore bottom, and straining to let go. Simply by looking at his tight features in his portraits you can easily visualize the man's intestinal distress. The 20th-century Freudian psychoanalysts had a field day analyzing the long-dead founder of Protestantism and his bowel movements. (Crucial historical note: Their own founder, Sigmund Freud, suffered from constipation as well.) A basic consensus is that Luther was neurotic, and his inability to move his bowels was at least partly psychosomatic in origin. Some Freudians such as Erik Erikson went so far as to place Luther's tight bowels as the underlying cause of the Reformation—i.e., angry, tormented, constipated man finds relief countering Catholic authority—which leads one to wonder what would have happened with modern laxatives. But who knows? Although Freudian psychoanalysis is widely discredited today, there is some biologic truth in psychological control of the orthosympathetic system, which, according to one scientific paper, makes the "colon longer, wider, drier, and sluggish."

Luther's Lavatory Located at Last

"THIS IS A GREAT FIND," SAYS STEFAN RHEIN, THE DIRECTOR OF the Luther Memorial, referring to a 450-year-old toilet. Not just any toilet, but apparently the one Luther actually sat on when working out his fundamental Protestant doctrines. Researchers excavating a garden on the grounds of Luther's house in Wittenberg, Germany, discovered an annex measuring 100 square feet—a room with a niche in one corner with what was unmistakably a toilet. Scholars say it is almost certainly the very toilet Luther sat upon during his 1517 religious epiphany. It seems to have been quite advanced for the times: It was made of stone blocks with a 12-inch seat with a hole, and, underneath, a cesspit connected to a drain. Luther's house apparently had a floor-heating system, which made it much more comfortable for long stays. Some questions remain (we know much less about daily life than we do about wars and philosophies): What about toilet paper? "We still don't know what was used for wiping in those days," explains theologian and Luther expert Martin Treu. Paper was too stiff—and far too expensive—putting to lie rumors that Luther used leaves from books written by his Catholic enemies.

And it was on one such sluggish day in 1517 that Luther had his famous religious epiphany, purportedly on the toilet, as many scholars believe. He was contemplating the New Testament book Romans (verse 1:17), when he suddenly realized that salvation comes by God's grace alone, not through human efforts such as buying indulgences. This became the central tenet of Luther's attacks on the Catholic Church and the theological basis of his eventual new Protestant Church, which also advocated a more personal relationship with God, one less dependent on clerics. Characteristically, Luther himself mentioned that his religious epiphany had indeed occurred on the cloaca (Latin for sewer): "This knowledge the Holy Spirit gave me on the cloaca in the Tower." (One caveat: Monks may have also used *cloaca* to refer to a warm room, and *in cloaca* to mean "in the dumps," as in depression, but it was far more widely used to mean sewer, latrine, privy, and the like.) He vividly described his immense relief, presumably after evacuating his bowels of feces and his mind of what he thought of as foul Catholic doctrines. In his words, he now "felt totally newborn, and through open gates I entered paradise." Some scholars now dispute Luther's own account of this cloacal epiphany and point out that his writings show a more gradual working out of Reformation doctrines. But Luther talked toilet, and so (at least traditionally) began the Protestant Reformation in this rather unlikely setting.

Of course the Reformation included other major founding figures such as John (Jean) Calvin and Huldrych Zwingli, leading most scholars to speculate that the time was right in general for a new form of Christianity. But Luther was the most prominent and certainly the most outspoken of Reformers, particularly with his, shall we say, frequent fecal fulminations. Yes, his 1517 religious epiphany was only one of many times Luther was potty-mouthed and potty-minded. He was not at all squeamish when it came to scatological references in his sermons, speech, and letters: "But I resist the devil, and often it is with a fart that I chase him away."

Speaking of the devil, the toilet was seen as a favorite haunt of his; Luther apparently thought of his constipated torments as devilishly inspired. A contemporary Reformation print, for example, shows the pope as having been born out of a she-devil's rectum. Times were more expressive then. Take Luther's evocative "Dear Devil . . . I have shat in my pants and breeches; hang them on your neck and wipe your mouth with them." Some churchmen objected to this sort of language. Sir Thomas More of England called Luther a "buffoon . . . [who will] carry nothing in his mouth other than cesspools, sewers, latrines, shit and dung . . ."

But Luther's other words and actions carried much weight, split the Western church, and changed Christianity. And in Luther's defense, his scatological mode of fighting evil is certainly invigorating, very direct, and definitely in keeping with his desire for a more "earthy" Christianity. As he once wrote: "If we once recognize Satan to be Satan, why, then it is easy enough to confound his pride by saying, 'Kiss my ass.'" (Presumably a putative devil would have had ample opportunity to do so, given Luther's own long sojourns on das Klo.)

Luther's Trials and Tribulations: The Devil's Punches?

MARTIN LUTHER WAS BY MOST ACCOUNTS A LOVING HUSBAND and a doting father, and a sincere if biased theologian, but his frequent mood changes have led some to speculate that he suffered from bipolar disorder. Others counter that with his list of bodily torments and afflictions, who wouldn't be a bit cranky and moody? (In fact, *cranky* comes from the German word *Krankenheit*, or illness.)

Luther's early portraits depict a gaunt, ascetic-looking person. (Luther claimed his early years as a monk ruined his digestion; his later Protestant portraits depict him as fat.) As diagnosed by doctors of the time, Luther suffered from bladder stones, chronic constipation, and hemorrhoids, and he ultimately died relatively young from coronary thrombosis.

But most interesting were his bodily battles in his later years with Satan, or so he said. The first attack occurred in 1527—Luther described tinnitus in his left ear that increased dramatically and extended into the left of his head, followed by roaring sounds, sickness, and vertigo. Finally he fell exhausted (but fully conscious) into his bed. When he woke the next day, most of the symptoms were gone or diminished, except for the tinnitus, which fairly continually plagued him the rest of his life. Satan or something else? Doctors now speculate that these attacks were actually manifestations of Meniere's disease, a chronic inflammation of the ears that produces tinnitus and vertigo, first described medically in the 1800s, long after Luther's death.

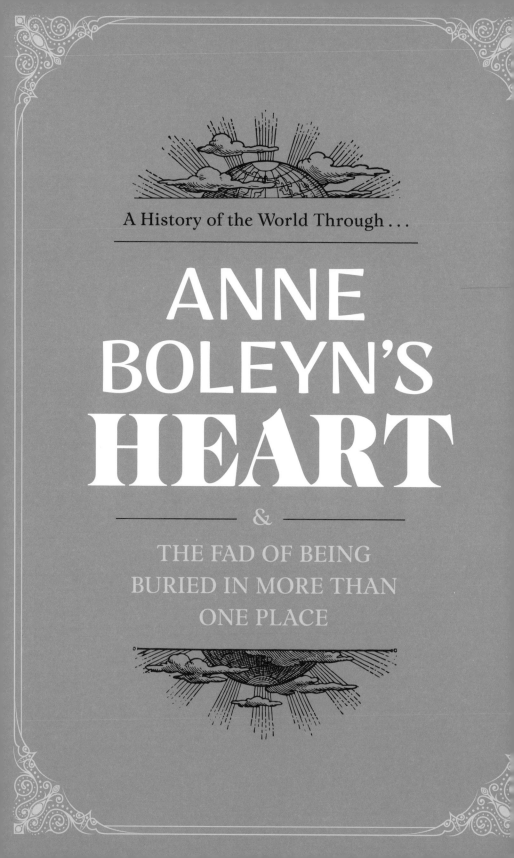

A History of the World Through . . .

ANNE BOLEYN'S HEART

&

THE FAD OF BEING BURIED IN MORE THAN ONE PLACE

(C. 1501–1536)

I F YOU ASKED PEOPLE WHAT body part you would associate with Anne Boleyn, most would probably say her head. Logical, of course. It was the part famously detached from the rest of her body on May 19, 1536, at the orders of her husband, English king Henry VIII. Henry had broken with the Roman Catholic Church to divorce his first wife, Catherine of Aragon, to marry Anne. But when she failed to give him the male heir he so desired, he decided to move on to another wife (Jane Seymour, wife #3 of his eventual six wives), accused Anne of treason and adultery (even with her brother), and ordered her execution.

But let's put Anne's head aside. It is her heart that concerns us here, a heart that, according to stories, was also separated from her body. Some say that Henry, still in love with her, requested that her heart be removed so he could keep it even while the rest of her body was buried without Christian rites in an unhallowed grave in the Tower of London. Others say that Anne made a final request that her heart be taken to the church in Erwarton, Suffolk, where she had been the happiest. No one is sure which story is true, or even whether her heart was taken out of her body at all, since it had been the rage in the past but wasn't very common by the time of Anne Boleyn's death.

Let's step back a bit to medieval times, when taking the heart from a body and burying it separately (or otherwise keeping it, in a decorative box or a bag) was de rigueur, particularly among the upper classes. It was just one aspect of what's called "dispersed burial." Since something that is dispersed is something spread over a wide area, you can rightly conclude that the only way to make a burial dispersed is to make the body dispersible—disassembled, if you will—i.e., cut into different parts so as to be able to inhabit a fair amount of real estate.

Disembowelment and embalming had been practiced since the 7th century in European countries, particularly north of the Alps, but the whole idea of burying a less-than-whole body really caught on in a big way during the Crusades, the wars fought between Christians and Muslims between 1096 and 1291. Because those who were killed on the battlefield were so far from home and usually in a hot climate, it was difficult if not impossible to transport the corpses home intact for burial in consecrated ground. The way around this? Take part of them home: bones, entrails, or, most popularly, hearts. Thus were Crusaders neatly and compactly returned to their native lands for burial.

Because most of the Crusaders tended to be aristocrats, dispersed burial became associated with the upper classes and it became desirable primarily due to snob appeal. People wanted to follow in the footsteps of the likes of Richard the Lionheart who requested a tripartite burial, with his heart buried in the cathedral of Rouen; his brain, blood, and entrails in Charroux; and whatever was left over buried by his family in Fontevrault. While most people weren't quite so dispersed, they liked the concept. It was a sign of status to have your heart buried in the family shrine while your body reposed in the local church. Speaking of local churches, they welcomed the practice wholeheartedly. It was a way of spreading the wealth, literally: When more than one religious establishment housed parts of a dead emperor, king, or lord, each got funding.

But the Church with a capital *C* in the person of Catholic pope Boniface VIII did not approve. Boniface thought it an abomination and wrote a bull, an official formal decree, banning it in September 1299. Perversely, the ban made dispersed burial even more desirable. Because a person needed dispensation from the pope, arranging one

The Rather Unfortunate Incident of the Sun King's Heart

LOUIS XIV'S HEART, ENSHRINED FOR 77 YEARS AT L'ÉGLISE SAINT Paul–Saint Louis, vanished during the French Revolution when the church was plundered. One story holds that it wound up as a family treasure of the Harcourts, landed gentry in England, where it met an ignominious fate. According to raconteur Augustus Hare, Dr. William Buckland (who harbored the unique ambition to eat every living thing he possibly could) went to the Harcourts for dinner, when a silver snuff box containing the heart—a gray thing much like a walnut—was being passed around. Hare wrote, "Upon seeing it . . . Buckland said, 'I have eaten many strange things, but have never eaten the heart of a king before,' and, before anyone could hinder him, he had gobbled it up, and the precious relic was lost forever." Some say that he actually popped it into his mouth to determine exactly what it was, as he did when he was examining rocks, and inadvertently ate it.

Frankenstein's Author and the Unplanned Unburied Heart

IT SOUNDS LIKE IT COULD HAVE BEEN PULLED FROM THE PAGES of a horror novel—which is fitting because it concerns Mary Shelley, author of the novel *Frankenstein*. In 1822, her husband, the romantic poet Percy Bysshe Shelley, died at age 29 in a freak sailing accident and was cremated on a makeshift funeral pyre. But—possibly due to calcification from tuberculosis—his heart wouldn't burn (although some people say it was probably his liver, not his heart). A friend plucked it from the flames, and Mary chose to keep the heart rather than having it buried with the rest of his body. She supposedly carried the calcified heart in a silk bag with her wherever she went. A year after she died in 1852, the heart was found in her desk, wrapped in the pages of one of his last poems, "Adonais." Forty-seven years later, it was finally buried, interred in the family vault with Mary and Percy's son Percy Florence Shelley.

became the ultimate 14th-century status symbol. Some took the concept of dispersal to the nth degree, like Cardinal Béranger Frédol, who, in 1308, got papal dispensation to have his body buried in as many different places as he wished. (Sadly for him, it appears that in spite of his grandiose and widespread plans, he was buried in only one piece and so in only one place. This is one of the drawbacks of after-death plans: You are not around to ensure that your arrangements are followed.)

So even though the church weighed in against it, the custom persisted, particularly among the bold-faced names of the Middle Ages. By the mid-14th century, Pope Clement VI pretty much bucked his predecessor Boniface's intention, and gave a sweeping dispensation allowing all the French royals to split up their bodies however they saw fit. It then became quite common to have a body interred in church for public veneration and a heart given to the deceased's family for a more intimate burial. The fad began fading out a bit in England after the 1400s, but dispersed burial, particularly heart burial, continued strong in France and Germany through the 1800s.

Which leads us back to Anne Boleyn and her heart. Since heart burial wasn't a terribly common thing anymore in England in the 1500s, there's a lot of debate about that particular organ. Was it removed from her body by her ex-husband as a sentimental keepsake? Was it removed and taken to Sussex in line with her last wishes? Or was it just left in her chest? No one is sure.

What we do know is that a small inscriptionless heart-shaped tin casket was found in the chancel wall when St. Mary's Church in Erwarton was undergoing renovations in the mid-1800s. The clerk of the church said that a story handed down through the generations held that Queen Anne's heart had been buried there according to her last wishes, and this must be it. The casket contained only dust (which could once have been her heart). That was good enough for the church. They reburied it under the organ and put up a small plaque marking it as the probable burial place of Anne Boleyn's heart. It has become popular with tourists, who, after seeing it, can go wet their whistle in town at the local pub, the Queen's Head.

A Few More Modern Heart Burials

WHILE THE NOTION OF HEART BURIAL BECAME PRETTY MUCH A dead issue for most people, it still has held allure for some through the ages. Among the more recent:

♥ Thomas Hardy, writer (died 1928): body in the poet's corner of Westminster Abbey, heart buried with his family in Dorset. (Note: Some stories say that it's actually an animal heart substitute, because the surgeon who cut out Hardy's heart stashed it in a cookie tin where his cat found and ate it.)

♥ Pierre de Coubertin, founder of the International Olympic Committee (died 1937): heart buried in Olympia, Greece

♥ Otto von Habsburg, archduke of Austria and royal prince of Hungary (died 2011): body buried in Vienna, heart in Hungary's Benedictine Abbey

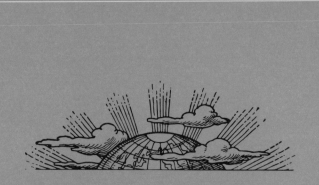

A History of the World Through . . .

Charles I's & Oliver Cromwell's Heads

&

TWO COMPETING IDEAS OF GOVERNMENT

&

THE RISE OF CONSTITUTIONAL MONARCHY

(1600–1649 and 1559–1658)

ET'S TALK ABOUT TWO SPECIAL human heads,
specifically those of two heads of state. One head was
removed while alive, the other, in a more macabre fashion, while dead. Better late than never, maybe.

The first severed head belonged to Charles I of England, who had
been a living king until a number of somewhat gloomy-minded Puritan
men, most notably Oliver Cromwell, decided it had to go. Cromwell
was a leader of the Parliamentary faction that wanted more power,
less taxation, and a more Christian nation. Once the king's problematic
head had been removed, their faction was quite successful (for a time).
Cromwell ruled in the king's stead as lord protector, and himself was
fortunate enough to keep his head attached and die a natural death.
But not long afterward, the (headless) king's son, Charles II, took
power, and then it was Cromwell's turn. His casket was opened, his
body dragged out, hung for a bit for effect, and then his dead head was
cut from his body and impaled on a stake near Parliament, serving as
a reminder of the perils of excessive political zeal.

And there we have it: beheadings as potent symbols of the literal
separation of the ruler from the body politic. Chop the ruler's head
off and you've really said—and done—it all. Cromwell's postmortem

beheading exemplifies the power of the symbol that extends even beyond the grave.

But beheading in the 1600s (and later) wasn't just about politics: Beheadings and other, often far more lurid public executions (think fun "entertainments" such as disemboweling, drawing and quartering, burning at the stake, and of course simple hangings) served as peculiar popular diversions throughout much of Europe (and elsewhere) until the 1800s.

If beheadings and the like were entertainment, what about those on the other side of the axe? Interestingly, the condemned often took on their expected role as committed actors in the drama; it was standard practice to face the axe (or the rope, etc.) with courage, dignity, or insouciance, and to make appropriate speeches to the crowd. Diaries of spectators to these bloody popular entertainments often admiringly record the behavior of the condemned much as we'd comment on a particularly good actor in a movie.

In the case of Charles I, he played the drama of his beheading to the hilt. In fact, he played it so well that it ultimately led to his son getting back the throne and subsequently removing not only Cromwell's head but also the living heads and other body parts of several of Cromwell's antiroyalist coconspirators.

First, what led to this rash of decapitations: Charles I had inherited the throne from his father, King James I. Sadly, Charles I had also apparently inherited his father's fascination with the concept of divine right of kings, which set out the not-then-unoriginal idea that kings are higher beings than others and that they ruled because of God. But the times were changing, and this idea was increasingly unpopular in England. Not surprisingly, a lot of the conflict centered around taxes: Charles I thought he had the divine right to levy them without parliamentary consent, a no-no to the rising middle class. It also didn't help that he had married a Roman Catholic. They were increasingly unpopular in Protestant England, and Charles soon became embroiled in all sorts of religious debates as well.

How to Behead People: The English Experience

IT'S NOT AS EASY AS IT SEEMS—IT TAKES A SKILLED WIELDER OF the blade. Beheading was usually reserved for nobility, and they weren't shy about wanting it done the right way: fast and with one blow. According to most scientists, a swift beheading is quite humane; estimates are that at maximum, seven seconds of consciousness might remain when the head is removed, but most believe the loss of consciousness is instantaneous. But there are executioners and there are executioners, and England had a long history of botched decapitators. Thomas Cromwell (the great-grandfather of Oliver Cromwell) suffered from one such executioner, who kept on hacking at his head until finally it fell off. An admiring onlooker stated: "So paciently suffered the stroke of the axe, by a ragged and Boocherly miser, whiche very ungoodly perfourmed the office."

Earlier, Anne Boleyn had anticipated this problem: For her beheading, she requested a skilled swordsman from France. Such are the prerogatives of royalty.

The bottom line was that revolts broke out, and at the end of it all, Charles was decisively defeated by the Parliamentarian general and member of Parliament Oliver Cromwell. Since Charles had already refused to accept the diminished role of a constitutional monarch, a number of more radical MPs, including Cromwell, decided that the head of state literally had to go. Charles was put on trial for treason.

Up to his trial, Charles hadn't behaved in a popular fashion—who likes arbitrary taxes and arrogant espousals of divinely sanctioned superiority? But if in life he wasn't an effective king, as he was facing death, he performed splendidly. He suddenly seemed to truly realize that Shakespearean idea of the world being a stage and that he was now its leading man.

During the trial, he put on an air of the calm acceptance of a martyr-to-be; his frustrated accusers in their harried public attempt to secure a conviction sometimes seemed more guilty than he. (And they may have had good cause for frustration; some modern accounts say that they gave Charles more than nine pretrial options to avoid death, and he turned them all down.) Charles was thus condemned, and his last days were spent refining his role as the superstar martyr. In the morning before his scheduled death, Charles and his courtiers undertook to complete the first version of a masterful book of propaganda, *Eikon Basilike: The Pourtrature of His Sacred Majestie in His Solitudes and Sufferings*, full of illustrations of the piety of the king and backhandedly condemning his accusers by oh-so-piously asking for their forgiveness.

On the day of execution, Charles played his last scene as solemn godly king before a huge crowd gathered to see this, the grand finale. He was led to a special scaffold draped in black, where he gave an eloquent speech (without his normal stutter), including the immortal lines: "I go from a corruptible to an incorruptible crown; where no disturbance can be, no disturbance in the world."

He then took off his cloak, gloves, and garter badge, lay down on the block, and gave the signal—and with a swift blow, his head came off. The audience reacted, not with cheers, as Cromwell and his Parliament had hoped. A young boy described how the blow of the axe was met with "such a groan as I have never heard before, and desire I may never hear again." Very soon, comparisons of Charles I and Jesus Christ were floating about.

For a while, it didn't much matter because Cromwell and his Parliamentarians had a firm grip—perhaps overly firm—on the new government. They outlawed the monarchy, but, also, in a rash of excessive Puritan fervor, most theaters and sports, and joyful Christmas celebrations (insisting the focus be on religion instead); they closed numerous inns, banned makeup, and established a "pious" dress code for women. Not a fun bunch. They basically set themselves up for a fall, which indeed came soon after Oliver Cromwell's natural death. The oldest surviving child of Charles I, the soon-to-be Charles II, was invited out of exile back to England, and after a spot of fighting with recalcitrant Parliamentarians, the monarchy was restored.

For generations thereafter, Charles I remained the poster boy for the just king facing unjust and rowdy democratically minded types—and also as a justification and a reason for keeping the English royalist. But, we must add that the Cromwellians were officially pardoned for daring to behead a king.

In 1985.

Tyburn Fair "Holiday"

MOST PEOPLE CAUGHT ON THE WRONG SIDE OF ENGLISH LAW— i.e., commoners—weren't beheaded but hung. But a variation of the same drama as that of Charles I often played out. The principal place of execution was Tyburn, near modern Marble Arch in today's London, and the days of execution were called Tyburn holidays. Batches of prisoners were driven in carts from the prison in Newgate to the Tyburn gallows, often watched by cheering throngs. The more insouciant the prisoners were, the better, and many, especially condemned highwaymen, dressed in their best and made clever and brave speeches. It didn't hurt that many prisoners had a bit of liquid fortification. The prison carts would usually stop at taverns, and the condemned men and women would sometimes get roaring drunk—a standard joke was that they'd pay for their drinks "when they came back."

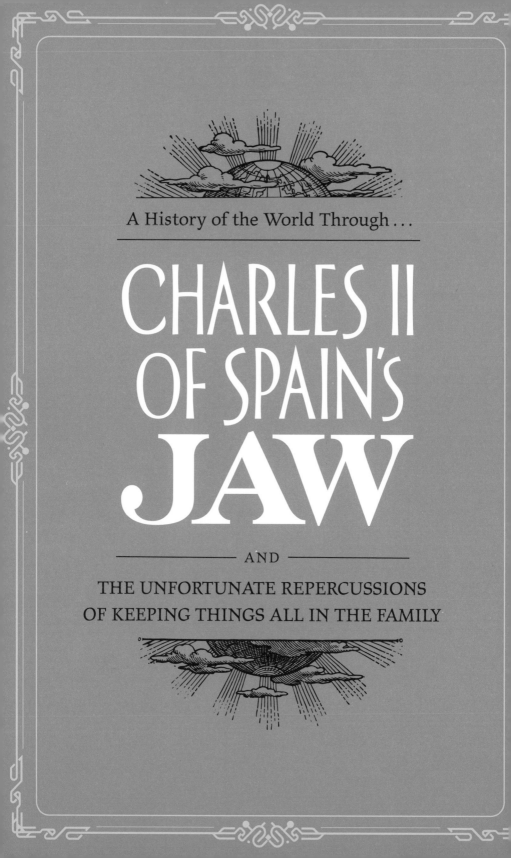

A History of the World Through...

CHARLES II OF SPAIN'S JAW

AND

THE UNFORTUNATE REPERCUSSIONS OF KEEPING THINGS ALL IN THE FAMILY

HARLES II OF SPAIN WAS a king who should have had everything going for him. He was a Habsburg, a member of the dynasty that ruled much of Europe, and had inherited all the perks that came with being the scion of a well-established royal family: money, land, power, and the throne. But he also inherited the family jaw, one that made him famous for all the wrong reasons. Charles's jaw distorted his face, made him drool copiously, and interfered with his speech. Known as the Habsburg jaw, it was the result of being part of a royal family that was *very* close—and it ended up being the symbol of the decay of their (Spanish) dynasty.

The Habsburgs were key players in Europe for more than 600 years—beginning in the late 1200s in Switzerland, moving into Austria, Germany, and Bohemia. In 1452, Frederick III became Holy Roman emperor, which cemented Habsburg influence. From that point on, Habsburgs were the leading dynasty in Europe, a status they maintained by strategic marriages to other powerful ruling families and, later, strategic marriages among themselves, something that led to problematic genes and, ultimately, the family jaw.

In scientific terms, the Habsburg jaw (sometimes called the Habsburg lip) is believed to be mandibular prognathism or maxillary retrognathism, with or without mandibular prognathism. Essentially, it's a jutting lower jaw. It is not, though, by any stretch of the imagination, what one might call a strong chin or a square jaw. Nor is it merely a prominent one. It's a jaw that greatly juts out, creating an underbite that makes it tough, if not impossible, to fully close the mouth. It is also coupled with maxillary deficiency—an underdevelopment of the upper jawbone, which makes the midface look dented, and often causes an abnormally thick bottom lip and tongue that make it difficult to speak clearly.

This genetic propensity was all fallout from the Habsburg method of keeping power: intermarriage. The Habsburgs initially gained power and new thrones, expanding their power base from Austria through much of Europe, through strategic marriages to other royal families. By the time they finally made their marital connection to Spain via a 1496 marriage of Philip I of Burgundy to Joanna of Castile, they had slightly changed their approach. They would maintain their power base through strategic marriage, but "strategic" now meant keeping it all in the family.

Intermarriage wasn't uncommon among other royal families, but the Habsburgs, particularly the Spanish branch, kicked it up a notch. From 1516 to 1700, the Spanish Habsburgs went in for marriages between close blood relatives (a.k.a. consanguineous marriage) nine out of eleven times, or nearly 82 percent of the time. Even though consanguineous marriage technically means marriage between second cousins or closer, the Habsburgs usually bypassed seconds. Instead, they typically opted for marriages between first cousins, double first cousins (which was more common than usual with them because of all that previous inbreeding), or uncles and nieces.

As such, Charles II's family tree was filled with very tortuous branches that doubled back on themselves: Charles's mother was his father's niece, making him both his father's great-nephew and his mother's first cousin, in addition to being their son; his grandmother was his aunt, so he was her nephew and grandson simultaneously; and

The Habsburg Jaw Lite

THE HABSBURG JAW GOT AROUND, NO QUESTION, EVEN WINDING up on the face of the much-celebrated Marie Antoinette, via her mother Maria Theresa of Austria. Marie had what could be called the Habsburg Jaw Lite—a slightly projecting lower lip that gave her a bit of a pout and was far from the disfiguring version that plagued other members of her far-reaching family. But it bothered her enough that she told artists to avoid painting her in profile.

all of his great-grandparents were descendants of the same couple, Philip I and Joanna.

With all this familial coziness, it's no wonder the Habsburgs, particularly the Spanish branch, wound up with some thorny genetic legacies. The jaw wasn't the only Habsburg sign of inbreeding. Most of them also had a Habsburg nose—with a bump at the bridge and an overhanging tip (the overhanging tip is another sign of maxillary deficiency)—not to mention nonanatomical unpleasantries, including tendencies toward gout, asthma, epilepsy, edema, and depression. No, it wasn't easy being a Habsburg, particularly as the interbreeding multiplied.

The hereditary jaw problem appeared in nine successive generations of Habsburgs. Some had it only slightly; others much worse. It's said that when the first Spanish Habsburg king Holy Roman emperor Charles V (a.k.a. Charles I of Spain) arrived in Spain from Ghent in 1516, a peasant jeered at him and his problematic physiognomy, saying, "Your Majesty, shut your mouth! The flies of this country are very insolent." It's unlikely Charles I reacted—he was the king, after all, who didn't need to follow orders from hoi polloi—but it's also unlikely he could have closed his mouth even if he tried.

Fast-forward 145 years from the first Spanish Habsburg to 1661 and the Habsburg jaw had reached its jaw-dropping apotheosis—or, rather, nadir—in Charles II. His jaw was so misaligned and his tongue and bottom lip so thick, that people couldn't understand his speech at all. There's a famous portrait of him by Baroque painter Juan Carreño de Miranda done in about 1685, when Charles was 24. In those days, artists made sure to placate their royal subjects by making them as good-looking as possible—in a kind of 17th-century version of Photoshopping—while still keeping the subject identifiable. As such, the portrait of Charles II does show quite the mighty underbite and a somewhat unusual jaw, but even so it's apparently not quite as distinctive as it was in real life. When the family of Marie Louise of Orléans was approached about a possible marriage with him, the French ambassador went to check things out at the Spanish court and wrote in a letter that "the Catholic king is so ugly as to cause fear." (Marie Louise had to marry him anyway. It was not a happy marriage.)

The jaw wasn't the only problem Charles had to deal with. He also had his not-so-parental parents to deal with. He wasn't raised in a loving and nurturing environment on any level. Because he had a number of health issues from birth, the sole focus was on keeping him alive to inherit and retain the Spanish throne for the Habsburgs. Nothing else mattered, not even little things like education, which, in the Habsburg perception, wasn't necessary. This meant Charles didn't get a formal education and was probably illiterate. (Some reports say he was mentally challenged; others that he simply appeared so because of his lack of schooling.) Even talking and walking weren't considered important. Charles reportedly didn't speak until he was four years old and his mother, wanting to keep him "healthy," didn't let him walk, and instead had him carried until he was eight or ten, depending on the source.

His adulthood wasn't much better than his childhood. He continued to have health problems and speech and personality issues. By the time he reached his 30s, his health grew worse. His hair fell out, he had difficulties walking, and he had hallucinations. He was once described as "short, lame, epileptic, senile and completely bald before 35, always on the verge of death but repeatedly baffling Christendom by continuing to live." It's no wonder that his subjects called him El Hechizado (the Hexed).

In the final irony, the Habsburgs lost Spain because of their efforts to keep it. Scientists from Spain's University of Santiago de Compostela who studied the Habsburgs concluded that all the inbreeding so damaged their genetic makeup that they were ultimately unable to reproduce. Charles II couldn't breed. His death in 1700 ended the Spanish Habsburg dynasty. Having no children, Charles named six-year-old Philip d'Anjou, grandson of his half-sister Maria Theresa and Louis XIV, as his heir, a move that ultimately led to the War of the Spanish Succession.

How A Certain Shade of Yellow Became the Habsburg Family Brand

MANY HABSBURG BUILDINGS, PARTICULARLY OFFICIAL ONES, from Vienna to Krakow to Cordoba, share a common color, a striking yellow-gold, called Habsburg yellow. The yellow color became inextricably tied to the Habsburgs, beginning when they took black and yellow as their family colors—the yellow because they held the title of Holy Roman emperor, which was featured on the official banner. In medieval times, yellow actually had negative connotations, associated with poison, jealousy, and deceit. But it was also the color of gold, thus wealth and power, and the positive attribute caught on.

So yellow became a quick "this is ours" identifier for the Habsburgs. Throughout their widespread multinational empire, they painted official buildings the same yellow, presenting a unified front. And its use began spreading beyond the Habsburgs. Because the Habsburgs were big on yellow, so too then were court hangers-on and bourgeois wanna-bes. In the 19th century, Habsburg yellow became the color of choice on upscale residences and villas, and eventually, in a trickle-down paint moment, on lower-class homes, farmhouses, and the like.

Habsburg yellow even pops up in the New World, due to a Habsburg strategic marriage. Habsburg daughter Archduchess Leopoldina married the Portuguese Dom Pedro, who became the founder and first ruler of the Empire of Brazil. As empress of Brazil, she introduced a little bit of her background to the country, including promoting the immigration of Austrians to her new home (some from the Tyrol valley founded the Brazilian village Tirol, where they still speak German) and putting a touch of Habsburg yellow on the Brazilian flag, where it still remains.

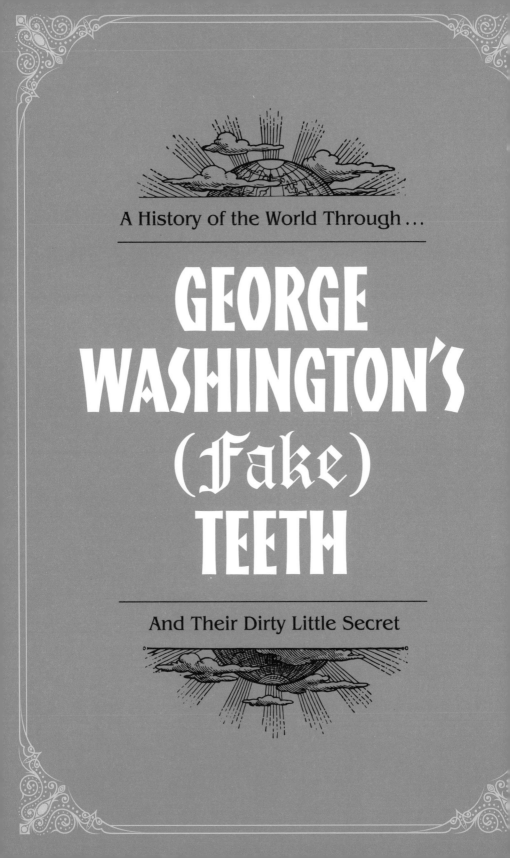

A History of the World Through...

GEORGE WASHINGTON'S (Fake) TEETH

And Their Dirty Little Secret

(1732–1799)

OUNDING FATHER GEORGE WASHINGTON HAD
a mother of a tooth problem. Rumor has long had it that
the peevish set of his mouth in the famous Gilbert Stuart
portrait of him was due to mouth pain from ill-fitting
false teeth. While we can't be sure about that, we know that he did
suffer from tooth problems, and that years after his death, the false
teeth became a pain not in his mouth, but in his legacy.

The dentures were the culmination of decades-long dental issues
that began when he was just in his 20s. At the age of 24, while commander of the Virginia Regiment in the French and Indian War, he
wrote in his diary that he had paid 5 shillings to "Doctr Watson" to have
a tooth pulled. The extraction was the first of many. As Washington's
public standing went up, his dental health went downhill. His diary
entries and letters from the battlefield make for exciting reading—for
a dentist. He regularly talked about toothaches, gum problems, and
dental treatments, while his ledger entries show purchases of toothbrushes, toothache medicine, and teeth scrapers. By the time he was
49 and commander of the Continental Army, Washington sported
partial dentures. When he hit age 57 and became the first president
of the newly formed United States, he was down to only one tooth

How Washington's Dental Issues Helped Bring Down Cornwallis

FOR ALL OF HIS WRITING ABOUT HIS TEETH IN HIS DIARY AND letters, Washington was pretty tight-lipped about his dental problems. So it's no wonder he was embarrassed when the British intercepted a mail packet of official military correspondence that included a letter to his dentist in Philadelphia asking him to send some dental cleaning tools to New York City (where the army was encamped). He needed them because he had "little prospect of being in Philadelphia soon" but wanted to keep his teeth healthy. As it turned out, having that letter intercepted (even if it embarrassed him) was a good thing indeed. The commander of the British army Sir Henry Clinton took the "I won't be in Philadelphia" comment to mean that the American and French troops would stay around New York City, so he didn't have to reinforce Cornwallis's troops in Yorktown. He was wrong. Washington and his French counterpart intended to march south to engage Cornwallis and his men, and they did just that—and defeated them.

of his own. But the adoring public wouldn't have noticed, since, like many other tooth-compromised people who could afford snappy new teeth, Washington opted for full dentures.

It was long said that the dentures were wooden, which led to speculations about splinters in the gums and talk about the need to keep said dentures nicely sanded. But the wooden teeth, like the "I chopped down the cherry tree with my little axe" story, were fiction. Yes, there were (and had been) dentures made of wood, but by the mid-1700s, most dentures—particularly those of the well-heeled if bad-toothed—weren't made of wood at all, but out of more toothlike material, including teeth themselves. As one would expect, a general, leading statesman, and wealthy plantation owner wouldn't have low-rent choppers, but the crème de la crème of false teeth available. That is just what Washington had: a denture plate made of carved ivory from hippopotamus tusks into which were fitted teeth—made of teeth. Some were parts of horse and donkey teeth cut to resemble human teeth, but most were actual human teeth. It is the latter point that has transformed a rather run-of-the-mill-for-the-times tooth problem into a modern ethical and public relations problem. At issue is the simple question of provenance: Just *whose* human teeth was Washington chewing with?

In the Mount Vernon ledger books, there is an entry dated May 8, 1784, a few months before Washington's dentist Dr. Jean Pierre Le Mayeur was to visit, for a payment of 6 pounds, 2 shillings "By Cash pd Negroes for 9 Teeth on Acct of Dr. Lemoire (sic)." It's unclear whether Washington purchased the nine teeth from the plantation's enslaved people for his dentist to use in his own dentures, or for his family, or simply as a gift to Le Mayeur. But it is clear that they were quite the bargain. Le Mayeur had advertised in New York that he would pay two guineas each for "good" front teeth. "Good" apparently didn't only mean "in good shape," but also white . . . and not as in enamel color. In a Richmond ad, he made it clear, stipulating that his purchase offer was "slaves excepted." The nine Mount Vernon teeth would have garnered over 18 pounds at white prices; since they belonged to nonwhites, they were less than a third of the price.

CONTINUED >

A Down and Dirty
History of Dentures

THE OLDEST FAKE TEETH, FOUND IN MEXICO, DATE BACK TO 2500 BCE and were fashioned of animal teeth, probably wolf. Animal teeth were pretty much the tooth of choice for centuries, or so we presume, because there weren't many other ancient examples found until 700 BCE. The Etruscans would take human or animal teeth, fit them into a gold band, and either bind them or pin them to existing teeth with metalwork, and voilà! Ancient bridgework! They had a tendency to deteriorate, so they needed to be replaced fairly often, one of the reasons they were mainly for the rich. But they worked well enough for them to be used for many centuries.

The next great leap in fake toothery came about in 16th-century Japan and the invention of wooden dentures. Artisans would carve tooth replacements to match wax impressions made of a person's mouth, making these tooth replacements a much more precise fit and a popular choice. Wooden dentures stayed in the false-tooth game until the beginning of the 20th century. They weren't the only choice, though; dentists still used animal teeth as well, and in the 1700s, ivory—from walrus, elephant, or hippopotamus tusks—became the newest innovation. Next, in the later 1700s, porcelain became the tooth rage. Called "incorruptable" dentures, they were hand painted by French dentists to make them look more natural. But while they looked good, they didn't work too well—porcelain breaks easily. And here's where things reverted a bit again, back to animal teeth and, the most desirable, human teeth.

Problem was, it was difficult to get human teeth on demand, particularly because most people presumably needed their teeth themselves. Most live people, that is. Dead people didn't need their teeth . . . which is why the dead became a major part of the tooth supply chain. It wasn't uncommon for dentists to harvest the teeth of corpses. Years after Washington, so-called Waterloo teeth were one of the more common types of dentures in France, so named because they came from the tens of thousands (estimated to be as many as 50,000) soldiers who were killed in the 1815 Battle of Waterloo. Executed criminals were also handy for teeth-farming. All well and good. But the need for human teeth was great enough that live people also entered into the supply chain. People plagued by poverty sometimes sold their teeth to get their hands on money. And then there were those people who became live teeth donors not because they chose to raise money in a rather desperate way, but because they didn't have a choice: the incarcerated and, as possibly with Washington, the enslaved. It was dentistry's not-so-hidden secret, the dark side of dentures.

Of course, it's highly probable that the "negroes" never saw the money paid for their teeth and that the money simply flowed back into Mount Vernon's income. Additionally, as Mount Vernon's historic preservation organization, the Mount Vernon Ladies' Association, neatly (under)states: "It is important to note that while Washington paid these enslaved people for their teeth, it does not mean they had a real option to refuse his request." Yes, even their teeth were subject to the whims of their toothless slave owner.

It wouldn't be surprising for the wealthy but penurious Washington to opt for the least costly teeth when it came to his dentures. He consistently tried to save on his teeth. Back in 1782, while in the field as army commander, he wrote to his manager (and distant cousin), Lund Washington, at Mount Vernon and asked him to look in a locked desk drawer where he had stashed a few of his own teeth after they had been pulled. He wanted Lund to send the teeth to him so they could be used in a set of dentures he was having made. So shelling out 6 pounds for "Negroes teeth" for his fake teeth rather than buying the more expensive ones fits his mold.

In spite of appearances and retroactive public relations problems with the possible provenance of his dentures, in his time, Washington was said to have been conflicted about the whole concept of slavery and even spoke about how much he would like it ended in the new republic. But this distaste for the institution didn't stop him from being a slave holder for 56 years (he had begun owning slaves at age 11). And just as he was quite pragmatic and penny-pinching when it came to his teeth, he was equally so when it came to his human property. Case in point: When Cornwallis surrendered at Yorktown, Washington made sure to reclaim all of the enslaved people from his plantation who had fled to the British during the war. (Jefferson did too.) He eventually freed them and all of the other Mount Vernon enslaved people, the only Founding Father with human property to do so. He had instructions in his 1799 will that everyone enslaved at Mount Vernon should be freed—but only after his death.

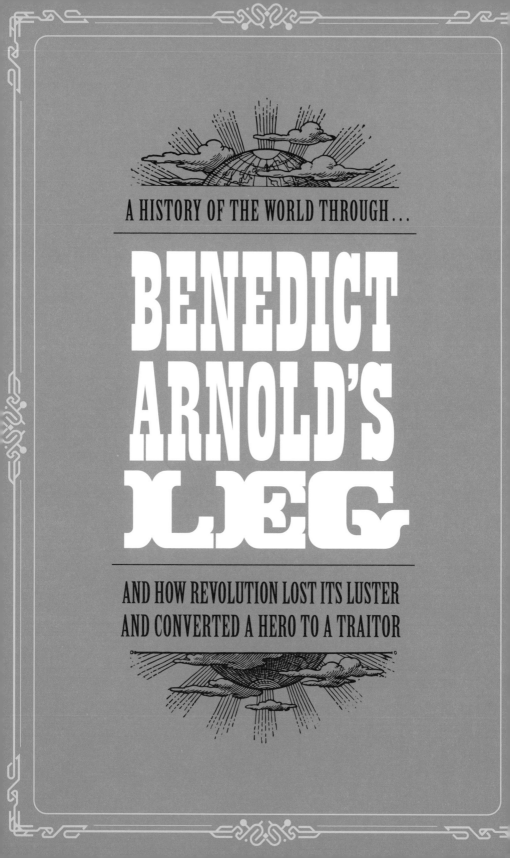

A HISTORY OF THE WORLD THROUGH...

BENEDICT ARNOLD'S LEG

AND HOW REVOLUTION LOST ITS LUSTER
AND CONVERTED A HERO TO A TRAITOR

(1741–1801)

 ENEDICT ARNOLD, THE REDCOAT TURNCOAT, is a staple of Revolutionary War lore, the first villain of the new union. (Of course, the British and the one-third of American colonists who were Loyalists would have probably disagreed. But then again, they lost the war.) Arnold was so hated that in the handwritten record of his birth in his hometown, Norwich, Connecticut, he is listed officially as "the traitor Benedict Arnold."

Yet in fairness to Arnold, he isn't considered to be 100 percent a traitor, just 90 percent. What's with the other 10 percent? Well, typically a human leg is about 10 percent of the body—and, unlike the man himself, Arnold's leg has gotten the hero treatment. It is immortalized in granite in New York's Saratoga National Historical Park, a monument bearing a bas relief of a boot and the following inscription:

In memory of the "most brilliant soldier" of the Continental Army who was desperately wounded on this spot the sally port of BUR-GOYNES GREAT WESTERN REDOUBT 7th October, 1777 winning for his countrymen the decisive battle of the American Revolution and for himself the rank of Major General.

The discerning reader will note that it doesn't mention the name of that "most brilliant soldier." Yes, even in a monument recognizing his heroism, Arnold remains the Revolutionary War version of He Who Shall Not Be Named. Yet former Civil War major general and military historian John Watts de Peyster felt strongly enough about Arnold's pre-traitorous actions that he had the nameless monument erected, a hat tip to a loyal leg.

That said, Arnold's leg didn't actually do anything particularly spectacular during any battle in the American Revolution other than get badly wounded. And, perversely, that wound probably contributed greatly to Arnold's dissatisfaction and eventual turnabout that made a traitor out of a hero.

Benedict Arnold most definitely had been a hero—from his first action leading a militia company in April 1775, to seizing British artillery stronghold Fort Ticonderoga just one month later, to leading a daring Canadian campaign to try to capture Quebec, in spite of the death of his wife, and managing to hold the British off even though his leg (*the* leg) was wounded. He was a bold-faced name of the American Revolution, called "the most enterprising and dangerous" field commander in the American army by British secretary of state Lord Germain.

His fighting career was put on hold after the Battle of Saratoga in 1777. He was leading a charge when he was wounded in his left thigh, the same leg that had been wounded in the campaign for Quebec, and pinned underneath his horse. He survived, and the Continental Army won the battle largely due to his leadership. As one of his soldiers later wrote, Arnold was "the very genius of war" on that day.

While soldiers might have respected his genius on the field, a recuperating Arnold felt that the powers that be in the Continental Army didn't. He was passed over for promotions, with five junior officers getting promoted ahead of him. He became convinced that other military luminaries (including Ethan Allen) were talking behind his back and taking credit for his work. And while stuck in Philadelphia with his leg healing, he began to feel unappreciated even by the public at large.

The Deadliest Enemy of the Continental Army

IT WAS DANGEROUS BEING IN THE CONTINENTAL ARMY, no question. But the British were far from the deadliest threat to the soldiers of the American Revolution. The worst killer facing the troops was smallpox. Smallpox swept through both Continental and British forces. But it affected the Continental troops much more than the British, who tended to be immune, due to inoculation or previous exposure to the disease, and who were quick to inoculate all soldiers when the smallpox epidemic spread. But Washington initially was loath to have the Continental troops inoculated, not wanting to risk having soldiers unable to fight for the time after being inoculated. So smallpox mowed down the troops. As the smallpox epidemic showed no signs of slowing (it lasted from 1775 to 1782), ultimately, Washington decided to opt for what became the first mass immunization policy in American history.

He wasn't alone in that feeling. Many other soldiers in the Continental Army, particularly non-officers, felt overlooked. Part of it could have been classism. Thousands of middle-class men were in militias, but few in the regular army. And as the war went on, fewer people, middle class or otherwise, actually *wanted* to be in the army. The rousing Spirit of '76 morphed into tired Spiritless '77. It was quite a contrast. At the beginning, everyone wanted to be part of the fight against the British. Just one week after Paul Revere did his "the British are coming!" thing on April 19, 1775, 16,000 men from the four New England colonies had formed a siege army. Two months later, in June, the Continental Congress took charge, making the New England force into the national Continental Army. Yet only six months later, in January 1776, General George Washington was already feeling a bit disheartened at the lack of men eager to enlist in the Continental Army.

As he wrote in a letter to Joseph Reed (more on him in a bit), "I no longer retain a hope of compleating the army by Voluntary Inlistements." Once the first "let's get the Redcoats" adrenaline rush passed, many of the colonists realized there were some unpleasantries about being in the army . . . like being wounded. Or killed. The army had to offer inducements to get people to enlist—cash upon signing, real estate, shortened service terms, or extended furloughs. And in 1777, when Congress said terms of service had to be at least three years or the duration of the war, whichever came first, they had to up their offers. Many "patriots" who could afford to stay away did just that, making the majority of soldiers young single poor men with no property, many in it for the perks instead of patriotic zeal. (As one soldier looked back on it, "As I must go, I might as well endeavor to get as much for my skin as I could.") Even so, the increased bounties, even when coupled with smooth-talking persuasive recruiters, didn't do the trick, forcing most states to switch over to conscription by the end of 1778.

Granted, Arnold took it a few steps further than the average disaffected Continental Army soldier. Some say he was spurred on by his young Loyalist wife; others held that he actually had been set up, and so forced, to a great degree, into treachery by Pennsylvania Supreme

Executive Council president Joseph Reed, who not only spread rumors that Arnold was involved in treasonous activities but also tried to prosecute him on treason charges that look a lot like setups.

Of course, only a few years later, Arnold actually did go the treason route, turning coat by planning to surrender West Point, which he was in command of, to the British. But the plot was uncovered, and while his coconspirator British major John André was hanged, Arnold wound up legging it, so to speak, to the British side. After fighting as a general for the British army in two battles, he ended up in England to finish his previously heroic life as an outcast, forgotten except as a traitor in the United States. But he did leave behind his leg legacy that was, to some degree, foretold.

Years before, when Arnold was leading British forces, he asked a captured captain in the Colonial Army what he thought the Americans would do to him if he were caught. The officer replied, "They would cut off the leg that was wounded at Saratoga and bury it with the honors of war, and the rest of you they would hang on a gibbet."

Pretty close.

How Much Did Benedict Arnold Make as a Traitor?: A Tally Sheet

£6,000 (INITIAL PAYMENT FOR HIS FAILED ATTEMPT ON WEST Point; if he'd have succeeded, he would have received £20,000)

£315 (for unnamed expenses)

£650 a year (for active service in the British army; paid until a treaty of peace was signed in 1783, then got £225 a year)

£2,000+ (share of the prize money when he and his men seized American shipping on the James River, Virginia)

In addition, his second wife was given a yearly pension of £500 by order of King George; and her children (including ones not yet born) were given a yearly pension of £80.

The upshot: Depending on which historian is doing the valuation, Arnold made between $55,000 and $120,000 in today's money for being a traitor. (Whichever amount it was, he wasn't happy with it. He tried to get more in 1785, asking for another £16,125, saying he needed it to cover the losses he incurred by switching sides. He didn't get it.)

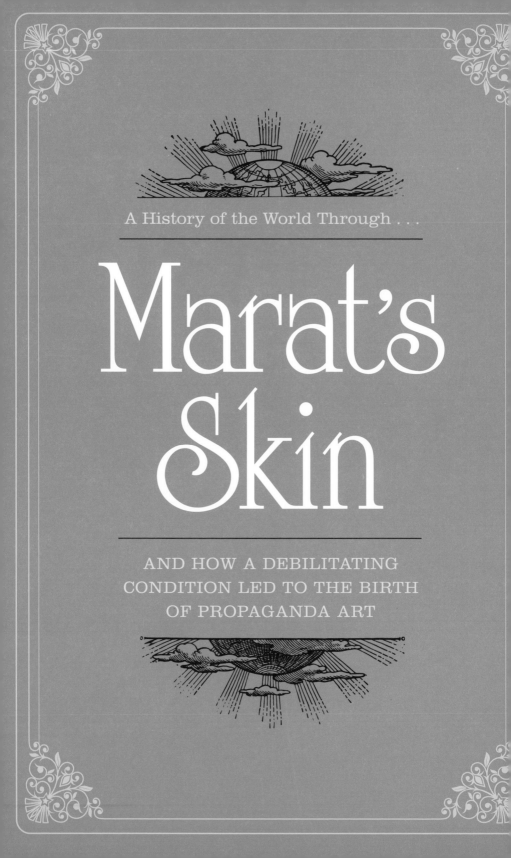

A History of the World Through . . .

Marat's Skin

AND HOW A DEBILITATING
CONDITION LED TO THE BIRTH
OF PROPAGANDA ART

NE OF THE MOST FAMOUS paintings in the world is of a man lying dead in a bathtub, where he had been soaking himself due to a painful skin condition. The painting is called *The Death of Marat*, and for a time it regalvanized the French Revolution and created a martyr. A cross between a CSI crime scene photo and religious art at its finest, it depicts French revolutionary Jean-Paul Marat just after he was stabbed by an assassin.

The assassination happened while Marat was working in his home "office," a bathtub with a board laid across it to hold his papers. It had been a difficult summer. His chronic skin condition had gotten worse, forcing him to resign from the ruling revolutionary government, the National Convention (which had very recently overthrown the French monarchy and beheaded the king), and drop out of the public eye. He was spending most of his time in his tub, the only place he could get relief from the burning and itching of his skin. Old friends and allies had been avoiding him. His radical hands-on—or actually heads-off— brand of revolution was going out of style.

From his tub, he was still writing letters to the Convention, hoping to help influence the new constitution, and still calling for action against the enemies of the people even though his former comrades, the Jacobins, were now moving away from the bloodshed he advocated. It seemed as though his days as one of the French Revolution's driving forces were over. After all, how could he make much of an impact if he was forced to soak in a tub rather than make speeches to rouse a crowd?

What Marat wasn't aware of was that he was going to have more of an impact on the French Revolution than he ever had and ever dreamed. And having to soak in a tub would actually be a big part of it.

On the evening of July 13, 1793, a young woman named Charlotte Corday came to see Marat in his tub, saying she had information about some escaped counterrevolutionaries, members of what Marat and his faction called the Girondists, moderates who had turned against the violence of the Revolution. Corday recited their names as Marat wrote them down. Then she suddenly whipped out a kitchen knife from her bodice and stabbed him in the chest, cutting Marat's carotid artery. Blood pulsed from the wound. "Aidez-moi, ma chère amie!" he called out to his wife in the next room, then slumped over in the tub and died.

Four days later, Corday was executed. She had had no chance to escape from Marat's home, but she hadn't wanted to anyway. A Girondist sympathizer herself, she had assassinated Marat because she felt he was leading France into civil war with his radical views. "I have killed one man to save a hundred thousand," she said at her trial. Killing him in his tub had been a last-minute change of plans. She had hoped to assassinate him in public so the crowd would then kill her—a sort of 18th-century twist on suicide by cop—and hoped that her suicide by mob would make her into a martyr for the Girondists. Instead, she had made Marat into the martyr and a major rallying point for the revolutionary Jacobins.

Marat hadn't seemed much like martyr material when he was alive. He had begun professionally as a doctor (to members of the French court, no less) and an Isaac Newton–refuting scientist who got miffed when the Academy of Sciences passed over him. He switched to politics,

So, What Happened to the Tub?

AFTER MARAT'S ASSASSINATION, THE TUB—SHAPED LIKE A high-button shoe and lined in copper—disappeared from his home, keeping the Revolution from a particularly potent symbol.

It's thought that his wife sold the deathbed tub to her neighbor, who in turn also sold it; it wound up in the possession of a Brittany parish priest in 1862. When an enterprising journalist from *Le Figaro* tracked it down, the priest realized he was sitting on something that could garner big money for his parish. He approached the Musée Carnavalet first but was turned down—the price was too high and they couldn't be absolutely sure it was Marat's tub. Next up was Madame Tussaud's wax museum, which offered 100,000 francs, but it's said that the priest's acceptance was lost in the mail and they lost interest. Finally, after rejecting other offers as too low (including one from P. T. Barnum), he sold the tub for a mere 5,000 francs to the Musée Grévin wax museum—where it plays a starring role in the (very lurid) re-creation of Marat's death.

becoming a full-time radical in 1788 at age 46 as the seeds of the French Revolution were being sown, and he quickly started making a name for himself as a voice for revolution. His impact grew after starting his own newspaper, *L'Ami du Peuple* (*The Friend of the People*) in September 1789. A strong supporter of the poor, he wrote incendiary pieces that roused the masses to fight for their rights. He was the revolutionary's revolutionary, excoriating not only the aristos, but also the revolutionaries who weren't revolutionary enough. And he was no softy. As he wrote in a July 1790 pamphlet, "C'en est fait de nous!" ("We're done for!") about counterrevolutionaries, "five or six hundred heads cut off would have assured your repose, freedom and happiness."

Preaching violent revolution comes with its drawbacks. Marat was forced to go into hiding frequently, often in the Paris catacombs and sewers. This wasn't the most pleasant place for anyone, but was particularly problematic for someone with a chronic and painful skin disease. Marat apparently suffered from a form of severe dermatosis that caused him to have intense itching and burning, as well as insomnia, insatiable thirst, headaches, and paranoia. Doctors and historians aren't sure whether the disease started because of his times in the sewer or whether that just exacerbated it, but they agree that it began sometime between 1788 and 1790 and that the only time he could get any relief was in the bath, which brings us back to July 13.

Marat's ignominious deathbed was elevated to a Gallic Golgotha by the Jacobins, a centerpiece of their swiftly devised "Let's Make Marat a Martyr" campaign. With Marat dead—and at the hands of an enemy Girondist, no less—the revolutionary Jacobins could now embrace him even though they had been distancing themselves from him before he was killed. A dead Marat was much more of a help to them than a live one; he was a lot quieter, for one thing. So they went to work establishing him as the first noble martyr of the cause: a revolutionary, anti-aristo, anti-religious messiah.

The leading painter of the time, who was, in effect, the official artist of the revolution, Jacques-Louis David was asked to organize his funeral and to commemorate the assassination on canvas. The funeral was a 6-hour affair that ended with a procession through the streets of Paris, punctuated by cannons booming every few minutes. Marat's remains were taken from their initial burial place to the Pantheon, resting place of the noted and revered, where he was eulogized by the Marquis de Sade (or Citizen Sade, as he preferred at that time), who pushed the messiah angle quite unequivocally: "Like Jesus, Marat loved ardently the people, and only them. Like Jesus, Marat hated kings, nobles, priests, rogues and, like Jesus, he never stopped fighting against these plagues of the people."

That was just the beginning of the Jesus connection. David's painting, completed in October 1793, showed the scene as it had been: board across the tub, papers and pen on the board, and Marat leaning out of the tub. But this Marat wasn't the middle-aged man with the pockmarked and lesioned skin that had brought him to the tub in the first place. Instead, he was a young man with glowing skin in a pose like that of Jesus in Caravaggio's *The Entombment of Christ*, or Michelangelo's *Pietà*. It was propaganda art of the first order, and at the orders of Robespierre and other leaders, David's students made several copies to spread the image . . . and the revolutionary word. More was to come. Busts of Marat took the place of saints and crucifixes in churches. Poems and plays were written about him. Marat had become the leading saint of the new nonreligious political quasi-religion in France.

But fame, even that of a revolutionary saint, is fleeting, especially when counterrevolutionaries counterrevolt. A little over a year after Marat's assassination, there was a coup d'état, the Thermidorian Reaction. The Jacobins were out, and Marat the Holy Martyr quickly became Marat the Reviled. By February 1795, he had been disinterred from the Pantheon. As a February 4, 1795, news story in *Le Moniteur Universel* reported, showing how far he had fallen in the public view, jeering children carried a bust of him around town, then, in Montmartre (which only a few months earlier had been renamed Mont Marat), shouted, "Marat, voilà ton Panthéon!" ("Here is your Pantheon!") and dumped it, ironically and fittingly, into a sewer.

So, What Happened to the Painting?

AFTER ROBESPIERRE WAS OVERTHROWN AND EXECUTED, *THE Death of Marat* was no longer the je ne se quoi of paintings it had been, although its fortunes would change. David asked that it be given back to him in 1795 when he was being prosecuted himself for his part in the Revolution. With his exile to Belgium, it pretty much disappeared from sight. In the mid-19th century, it was rediscovered and repopularized, most notably by poet/critic Charles Baudelaire. From there it went on to inspire countless other artists, among them Picasso and Munch, as well as poets and writers, including the classic play *Marat/Sade*.

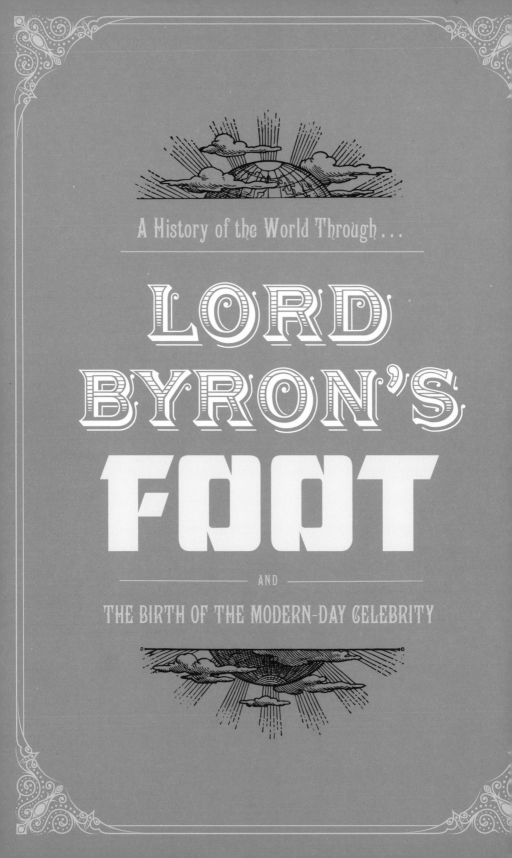

A History of the World Through . . .

LORD BYRON'S FOOT

AND

THE BIRTH OF THE MODERN-DAY CELEBRITY

(1788–1824)

 HE WALKS IN BEAUTY LIKE the night." So wrote English poet George Gordon, Lord Byron. But the hottest literary celebrity and heartthrob of his times, not to mention the principal founder of the Romantic movement, would have had a lot of trouble gliding alongside his lovely muse.

The problem: Lord Byron was born with a congenital foot disability (thought to be clubfoot by many) that made walking difficult. Edward Trelawny, a close friend of Byron's, described his unusual gait: "He entered a room with a sort of run, as if he could not stop, then planted his best leg well forward, throwing back his body to keep his balance." It embarrassed him terribly. Byron supposedly vowed to never let anyone see his foot, and his many mistresses described how he would surreptitiously leave their beds before morning.

How much this affected his creative output we can only guess. Many famous or prominent people have been born with so-called clubfoot (shortened tendons in the lower leg that force the foot to twist inward), from the ancient Egyptian pharaoh Siptah (and maybe King Tut) to writer Sir Walter Scott, from the notorious Nazi propagandist Joseph Goebbels to actor Dudley Moore and football player Troy Aikman, and many seem to have been relatively unaffected emotionally. But with

Byron it seems to have been another matter entirely, either due to its severity or to his own temperament. His acid-tongued contemporary essayist William Hazlitt once cruelly stated that Byron's "misshapen feet . . . made him write verses in revenge." Byron seems to have at least partially agreed: He once wrote in his unfinished and partially autobiographical play *The Deformed Transformed* (1824) how the main character, with kyphosis and "a cloven hoof," was driven to create due to his disability: "There is / A spur in its half movements, to become/ All that others cannot."

His self-styled "cloven hoof" served as a creative goad—and as a source of deep distress throughout his life. Lady Blessington, a friend and early biographer, said Byron was haunted by the taunts about his limping from his schoolmates at Harrow. Worse, Byron once said that his mother called him a "lame brat," maybe more than once. Byron often called himself "le diable boiteux" (French for "the limping devil") and said he simply could not "conquer the corroding bitterness that deformity engenders in the mind, and which, while preying on itself, sours one towards all the world."

But Byron at his sourest is the rest of us at our most sublime. In his short life of 36 years, he not only wrote classic poetry, including *Don Juan, Childe Harold's Pilgrimage,* and *Hebrew Melodies,* but he was also a principal in the Romantic movement that reveled in the dark romance of the medieval, the Gothic, and the preindustrial. In the House of Lords, Byron championed the cause of the poor; in Greece, he took on their cause of freedom against the Turks (and died tragically there). He also virtually single-handedly created a philosophical genre that placed solitary Person against the Universe, a philosophical take on the emotional aspects of Romanticism. True to form, he lived out his own tragic philosophy.

Lord Byron used his own contradictions—startlingly handsome and yet with a quasi-secret disability—to become the classic celebrity spokesperson (a title not yet invented) of the newly created Romantic movement. He was a true pioneer of superstardom, arguably the first to create and control a literary public image that even has a name:

The Medical Mystery of Byron's Foot

BYRON'S "SOUL CORRODING" DISABILITY SOUNDS FAR MORE intense than simple clubfoot, which, even in the late 1700s, was often fairly treatable by braces, stretching, and casts. So why didn't Byron achieve even a partial cure?

In 1959, a prominent British physician, Dr. Denis Browne, examined a special shoe and a strap-on calf legging worn by Byron, both of which were left in the possession of the descendant of Byron's publisher. Byron's shoe was long and thin, and his calf legging was unusually thick. Dr. Brown concluded that instead of clubfoot, Byron had a dysplasia, a failure of his foot and calf muscle to grow sufficiently. In other words, in addition to his tendons, Byron's foot and leg were also much smaller and narrower than normal. In fact, Byron's calf was probably "grotesquely thin" (doctor's words), so the thick padded calf legging would give the appearance of a normal calf under trousers. (According to some accounts, the sensitive Byron even wore this calf-enhancer while swimming.) As for the foot, "feet of this dysplastic kind are always stiff, so there would be a lack of ankle movement, which would account for the sliding gait described by one of the few accurate observers."

the Byronic hero. He had all the attributes of his creation: extremely talented, wealthy, mercurial, noble, tormented by a secret tragedy, and, of course, appropriately good-looking. Fellow poet Samuel Taylor Coleridge once wrote that Byron's face was "so beautiful, a countenance I scarcely ever saw" . . . with "eyes the open portals of the sun . . ."

Byron clearly knew the impact of that handsome face. He artfully controlled any artistic or published depictions of it. In an age before paparazzi, it wasn't all that difficult. He instructed his publisher John Murray to destroy any unflattering painted or engraved images while endorsing those he himself liked, not to mention commissioning particularly flattering portraits. Most notably, there is the famous work by artist John Phillips of Byron in Albanian kilt, embroidered cloak, and velvet waistcoat—he had bought several of these rather flashy outfits on a European tour—in which Byron looks like an exceptionally romantic brigand, though less a real brigand than a Hollywood fantasy brigand.

Like any good Hollywood star playing the part, Byron dieted and even used purgatives to keep his weight down. (He had a tendency to pudginess that he fiercely fought.) Meanwhile, he kept his prodigiously large and quite varied fan base engaged. As your quintessential brooding, handsome romantic poet, he predictably had a large number of female admirers, many eagerly seeking his autograph, a lock of his hair, or, best of all, a clandestine romantic tryst . . . which he was not loath to indulge. He was the classic "bad boy"—in the words of one mistress, "mad, bad, and terrible to know."

Byron kept many of the letters his adoring fans sent him. Like any card-carrying celebrity, he probably enjoyed the adulation, something suggested in a 2008 study of Byron fan letters by an Oxford historian. The letters themselves are often quite sexual and sometimes (theoretically) poetic, as with one fan who "trembled" as she gazed at his portrait and wrote these not-so-immortal lines:

Why, did my breast with rapture glow? Thy talents to admire?
Why, as I read, my bosom felt? Enthusiastic fire.

Some fans sought to "heal" Byron's wounded heart, seeing in him a "kindred spirit." And Byron ate it all up. There was a practical side to it too: They were, in effect, focus group data. By reading them, he could track fan reactions and his reception by readers. (He was also an avid reader of critical reviews, and had his publisher send any to him when he was abroad.) And while he often played the aloof romantic hero, the fact that he kept so many of his fan letters (although fans often requested he burn the letters after reading, as they feared their intimate nature) also suggests that he reacted to them even in his own writing—to some extent, he wrote what they wanted to read.

But for all the trysting with female admirers and saving of their letters, Byron's passions inclined him more toward males, particularly adolescents, something his executor and publishers assiduously sought to conceal after his death. Case in point: A love poem to a young Greek boy was said to be a "mere poetical scherzo." Yeah, right. One wonders whether Byron cultivated the "bad boy with women" trope more for a positive celebrity image than out of strong desire. (He apparently loved dressing his female companions as pageboys, which certainly suggests something.) After all, same-sex sexuality was illegal in the sexually oppressive Britain of the 1800s. He was finally forced to leave due to certain sexual "excesses," including, according to the written records of the time, the immoral act of "." [sic] (Yes, they couldn't dare mention same-sex relations, so instead they wrote ".")

It's the darker side to the original Byronic hero's tale. Approbation for his sexual preferences and that ever-present "soul crushing" disability, as he called it, made it hard for him to simply be happy, be himself, and enjoy his fame and literary prowess. He grew prone to rages, debaucheries, drunken excesses, and depressed agonies. Modern psychologists call all these classic signs "body dysmorphia," the mental disorder characterized by excessive brooding on one's bodily imperfections. Granted, with all that brooding came some of the world's greatest poetic masterpieces—but sadly, for Byron, it was not enough.

Byron: The First
Count Dracula?

IT WAS A DARK AND STORMY NIGHT . . . OR RATHER
several nights . . .

And Lord Byron, his new friend poet Percy Bysshe Shelley, Shelley's
fiancée Mary Wollstonecroft Godwin, her stepsister Claire Clairmont (who
had once been involved with Byron), and Byron's personal physician John
Polidori were spending time at Byron's rental home near Lake Geneva
during an uncharacteristically dark, stormy, and cold June. There was an
atmosphere of sexual tension at the villa: Polidori was interested in Mary
(who didn't reciprocate) and Claire in Byron (who did reciprocate, but
only half-heartedly), and Shelley was getting more and more agitated.
Byron suggested they all tell or write ghost stories to pass the time, and,
for several nights thereafter, they read aloud ghost stories and poems.
There, Mary, after a nightmare, came up with her immortal novel *Frankenstein*; Byron introduced his "Fragment," one of the earliest vampire
stories, and Polidori was inspired by Byron's piece to eventually write
his own vampire novel, appropriately titled "The Vampyre," featuring
vampire Lord Ruthven, who bears a distinct likeness to Byron himself.
This was the first appearance of the romantic, aristocratic vampire. Years
later, a young Bram Stoker was inspired to create a literary vampire—Lord
Ruthvn merged with Vlad the Impaler, a.k.a. the immortal Count Dracula.

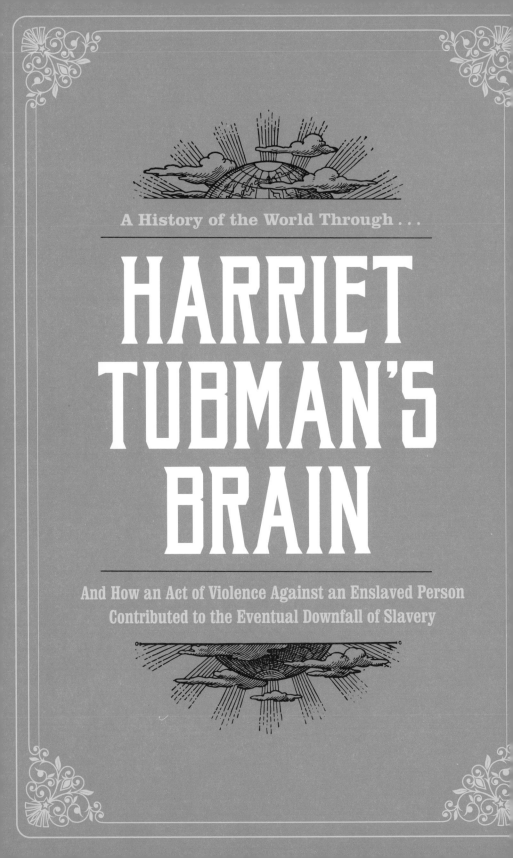

A History of the World Through . . .

HARRIET TUBMAN'S BRAIN

And How an Act of Violence Against an Enslaved Person Contributed to the Eventual Downfall of Slavery

(C. 1822–1913)

 OMETIME IN THE MID-1830s, an enslaved teen-
aged girl called Minty refused to help an overseer who
was trying to catch a fugitive who had run away from
the fields. The infuriated overseer picked up a heavy
iron weight and threw it, hitting Minty with full force. She recalled:

> The weight broke my skull and cut a piece of that shawl clean off and
> drove it into my head. They carried me to the house all bleeding and
> fainting. I had no bed, no place to lie down on at all, and they laid
> me on the seat of the loom, and I stayed there all day and the next
> . . . [Then] I went to work again and there I worked with the blood
> and sweat rolling down my face till I couldn't see.

A not-untypical event in the life of an enslaved person in the pre–
Civil War South, but one with an atypical result: The brain injury
helped make this young girl into the iconic woman we know today
as Harriet Tubman, famed conductor of the Underground Railroad.

The hurled weight caused Tubman's skull to fracture. According
to modern neurologists, that fracture brought on emotionally medi-
ated synesthesia, "a perceptual phenomenon in which stimulation

of one sensory or cognitive pathway leads to automatic, involuntary experiences in a second sensory or cognitive pathway." To put it in more digestible, nontechnical words, Harriet Tubman started hearing and seeing visions—while still maintaining her sanity. These both intensified her faith and convinced her she had supernatural powers, including the ability to see the future—which aided her in becoming the so-called Moses of her people.

Specifically, Tubman's brain injury caused her to get crushing headaches, narcolepsy, vivid dreams, and, most important, what neurologists now believe was "absence epilepsy," seizures that cause lapses in awareness lasting a few seconds. Tubman herself described it more simply. She said that she saw bright lights and colorful auras and heard disembodied voices, some of which she thought came from God. She considered them religious experiences; they, along with her dreams, were a trade-off for the recurrent pain she suffered.

Until her skull fracture, her life had been the sad norm for those enslaved. She was born Araminta Ross to enslaved cook Harriet Green and skilled enslaved (and later free) woodsman Ben Ross in Maryland. As with so many other enslaved children, she was put to work early. At only age 5 she was hired out to take care of a baby at night to make sure it stayed quiet, having to continuously rock a cradle or hold the baby in her arms. (If the baby cried, the mistress of the house would come and whip her. Tubman learned to wear multiple layers of clothing to avoid the stings of the lashings.) At age 8, she was hired out to wade in waist-deep water and collect muskrats from traps, then was switched back to domestic work. (She later said she preferred work outdoors for one key reason: It kept her away from the whippings of the exacting mistresses of the house.) And sometime between age 12 and 14 came the life-altering skull-shattering incident.

Fast-forward to 1849, when Tubman, then in her 20s, faced another life change, the death of her master. Tubman had since married a free Black man, John Tubman (and had changed her first name to Harriet), but was still enslaved. With her master's death she faced the possibility of being sold, separating her from her husband as well as her parents

Getting Hit on the Head and Becoming a Genius: Acquired Savant Syndrome

SOME NEUROLOGISTS HAVE CALLED THE EFFECTS OF TUBMAN'S injury a case of acquired savant syndrome, in which a brain trauma induces special talents. Savant syndrome itself, such as that which occurs with those born with autism and other early neurological conditions, is very rare, with a frequency of about one in one million. Sudden acquired savant syndrome is even rarer—only about 50 cases have thus far been documented. It usually occurs after a traumatic brain injury but can also occur post-stroke. One well-documented case cited by the Wisconsin Medical Society occurred after a 10-year-old boy was knocked unconscious by a baseball. When he came to, his brain had changed and he found he had acquired some amazing new abilities. Among other new talents, he could suddenly and easily do remarkable calendrical calculations—literally in seconds. He could quickly name the day of the week corresponding to a given date without needing to look at a calendar.

and siblings. She also knew that if she had a child, it would be the property of whoever bought her per the law at that time. She decided it was time to escape. (Her husband decided to stay.)

This was Tubman's introduction to the Underground Railroad, the network of homes and people who sheltered enslaved people on their way to freedom in the North. She made her way to the relative safety of Philadelphia (slave catchers were not remiss in snatching escapees and free Black people from non-slave states). So began Tubman's new vocation: guiding others to freedom, impelled and aided by her synesthetic religious visions. Over the next decade, she made numerous secret trips to Maryland and Delaware, freeing many people—some sources say 300, others 70; no one is sure of the exact number since her life became embellished with legend and myth. But the basic fact is clear: Enslaved people made it to freedom because of Tubman, some, including her parents and brothers, freed directly by her; others indirectly, by following her instructions and routes to safe houses; and others, inspired by her example.

Tubman credited her visions and dreams for her remarkable success as an Underground Railroad conductor. Because of them, she just "knew" which routes to take and which ones to steer clear of, and it just . . . worked. As she put it, "I was the conductor of the Underground Railroad for eight years, and I can say what most conductors can't say—I never ran my train off the track and I never lost a passenger."

This was no small feat in the antebellum United States, which was crawling with armed slave hunters seeking high bounties for escapees. Later, she used her powers to help the Union Army as a spy and scout, impressing the officers with her prodigious memory and geographical savvy.

After the war, Tubman settled in Upstate New York, where she farmed with her new husband. She became a pioneering woman's suffragist and opened a home for poor Black elderly people. As she aged, she grew progressively weaker, but, according to her grandniece:

It is said that on the day of her death, her strength returned to her. She arose from her bed with little assistance, ate heartily, walked about the rooms of the Old Ladies' Home, which she liked so much, and then went back to bed and her final rest. Whether this is true or not, it is typical of her. She believed in mind [over] matter. Regardless of how impossible a task might seem, if it were her task she tackled it with a determination to win.

Life or Legend?: The Problem with "Historical" Sources

AS WITH MANY ICONIC HISTORICAL FIGURES, THERE'S A GREAT
deal of myth and questions of accuracy surrounding Harriet Tubman's
life. African American studies professor and biographer Milton Sernett
calls her "America's most malleable icon" because stories about her
have often been changed by those wishing to enlist her memory to
their cause. For example, during the 2008 Democratic presidential
primary, a writer condemned women supporting Obama over Clinton, citing a quote by Tubman: "I could have saved thousands—if
only I'd been able to convince them they were slaves." Nice line,
but it looks like Tubman never said it. Sernett thinks it came from a
fictionalized biography.

Add to the problem of separating fact from fiction is the fact (true)
that the illiterate Tubman never wrote an autobiography, so many
accounts of her life are marked by the unconscious and conscious
bias of those who wrote about her, even in contemporary accounts.

Take a key main source about her life, the 1886 book *Harriet Tubman: The Moses of Her People*, which is based on stories Tubman told the writer Sarah Bradford. But Bradford, who was white, apparently censored some of these reminiscences, particularly in the last (and definitive) edition. According to an article by Professor Jean Humez, Bradford took out accounts of Tubman teasing her slave master (singing songs like "I'm Bound for the Promised Land," confident that he wouldn't understand their underlying meaning that she was going to escape). Bradford's censorship was probably well intentioned; she was anxious to win over white readers in the context of the racist Reconstruction era, but she doesn't give us a complete picture of Tubman. Bradford omitted this from the last edition as well: "Tubman on the stage version of *Uncle Tom's Cabin*: 'I've heard *Uncle Tom's Cabin* read, and I tell you Mrs. Stowe's pen hasn't begun to paint what slavery is as I have seen it at the far South.'"

A History of the World Through...

THE BELL FAMILY'S

EARS

And an Obsession with Hearing
—and Not Hearing

—————— OR ——————

A Quest for Visible Speech and
What It Gave the World

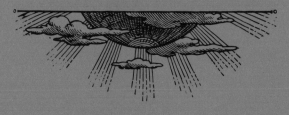

(1847–1922)

ET READY FOR AN EARFUL. We're talking ears, historical ears—those of inventor Alexander Graham Bell and of his father, grandfather, mother, and wife, not to mention one belonging to a corpse—all leading to that important hearing (and speaking) invention, the telephone.

We begin with the ears of Alexander Graham Bell's grandfather, Alexander Bell. You might say listening to sounds (and then reproducing them) was the Bell family business. Grandfather Alexander was a pioneering authority on phonetics and elocution, and passionate about speech: "Perhaps, in no higher respect has man been created in the image of his Maker." Others would passionately disagree.

Grandpa Bell passed on his fascination with human sound to his son (and for many years, his assistant), Alexander Melville Bell. Melville became an eminent specialist in elocution at the Universities of Edinburgh and London, where he developed Bell's Visible Speech, a written system for transcribing and reproducing spoken sounds. It was an elegant and seemingly useful system with symbols indicating precisely how to position the throat, tongue, and lips to make any given sound, and was especially designed for deaf people who wished to speak, since they could not hear the sounds they were making.

Bell's Visible Speech was a major entry into the growing debate of the times that continues to this day: Do we teach deaf children to speak, or to use sign language, or both? In the early 1800s, the United States tended to emphasize sign language. In 1817, what would become the American School for the Deaf was founded in West Hartford, Connecticut, and American Sign Language (ASL) was developed. But by the mid-1800s attitudes were changing, and the idea that deaf people should focus on speech rather than sign language was gaining prominence. The Bells were front and center in this increasingly acrimonious debate of talking versus signing—and along the way, one of them would invent the speaking machine par excellence: the telephone.

Enter a new set of Bell ears. In 1844, Melville married one Eliza Symonds, who soon became the mother of Alexander Graham Bell (yep, that one). Ma Bell influenced her son very directly—as a hearing test subject. She had developed profound hearing difficulties and was able to only partially hear with the crude aid of a hearing trumpet. Alexander began seeking methods to help his mother experience sounds more clearly. He would have his mother press an ear close to the piano when he played, and would speak directly to the crown of her forehead, creating an echo chamber in her skull. Clearly, Alexander had a telephonic (far-hearing) career ahead of him.

Father Melville kept doggedly promoting Bell's Visible Speech, first in England, then in North America, where he moved with his family. With him (now in body and in spirit) was son Alexander, a fully fledged speech elocutionist in his own right. Both father and son were clearly on the side of spoken communication rather than signing for deaf people.

Meanwhile, the controversy about how deaf people should communicate grew—with some ugly aspects. The United States was experiencing an unprecedented wave of immigration from non-English-speaking areas and some people were getting fixated on assimilation. To them, ASL was, in effect, a non-English language because it relied on gestures and hand movements—it was "un-American," so it had to go. ASL actually began to (temporarily) fade a bit, even while societies of

Controversy: Who *Really* Invented the Telephone?

ANTONIO MEUCCI, OF COURSE. AT LEAST ACCORDING TO THE Italian government, which gave him the unequivocal title "Official Inventor of the Telephone."

Meucci first developed an acoustic telephone (as in ship's speaking tubes) for the stage in 1834, then, after migrating to New York, began working on the idea of electrical current alterations to reproduce sound. Meucci filed a "patent caveat" (sort of a preliminary patent) in 1871, but probably due to money concerns, he never proceeded further. Frustratingly for him, although he claimed to have described his invention in a New York Italian newspaper, *L'Eco d'Italia*, in 1861, all issues of that edition were lost.

After Bell's later success, Meucci sued the Bell company; the case went all the way to the Supreme Court but was dismissed. Meucci wasn't the only one suing—the Bells faced a mountain of lawsuits in the early years, suggesting that, as is often the case, when the time is right, many people have the same general idea.

deaf people still liked it. The acrimony got occasionally fierce. There was the taint of eugenics in this debate, with worries about hereditary deafness, deaf people breeding with deaf people. The ugly anti-ASL idea was clear: "Superior" people talked.

Son Alexander certainly practiced what he preached. He went to Boston, where he began teaching elocution to those who were deaf and where he also started tutoring a wealthy young deaf student, Mabel Hubbard. Hubbard's father had financed the first school for deaf people that promoted orality, with daughter Mabel as his inspiration. Rendered completely deaf at five years old by a bout of scarlet fever, she had learned to fluently lip-read and speak not only English but several other languages as well. Bell was smitten by this confident and highly intelligent young woman, who initially didn't reciprocate. She thought Bell was interesting but "he had a horrible shiny [hat]—expensive but fashionable—and which made his jet-black hair look shiny. Altogether I did not think him exactly a gentleman." Nevertheless, the two got married. Mabel Bell would contribute greatly to Alexander's career and to the development of telephony.

During his off-hours from teaching speech to deaf people, Bell experimented with sound, basing some of his early work on a "phonautograph," which translated sound via electrical impulses into visual inscriptions. Bell made a unique and rather macabre "dead ear phonautograph" using the ear (as well as a section of the skull) of a dead man. He attached a recording stylus to the ear; the ear bones vibrated upon encountering sound, and the stylus (a piece of straw) then traced the patterns onto a coated glass plate. Bell's initial idea was not what you might think—he had those who were deaf on his mind, not those who could hear. He hoped to create a machine that would help deaf people visualize speech—a sort of *electric* Bell's Visible Speech.

The idea quickly grew into a more general one when Bell realized that "it would be possible to transmit sounds of any sort if we could only occasion a variation in the intensity of the current exactly like that occurring in the density of air while a given sound is made." Hmm, an electric version of Bell's Visible Speech that actually reproduced all the sound movements with electricity instead of human mouths and that, instead of on a plate, rendered the sounds into the air . . . and into the (hearing) ears of a listener. A kind of *Bell's Visible Speech that talked.*

Bell was clearly on to something. In 1876, he patented his "electrical speech machine"—which fortunately did not use dead human ears and was soon called by the far more euphonious word *telephone.* The first intelligible words spoken on the telephone (by Bell to his assistant) were the now famous but ineloquent "Mr. Watson—Come here—I want to see you," but they did the job. (Some time later, Bell made the first long-distance call—over a distance of 4 miles.) Mabel Bell then entered the picture in a seminal way. Alexander did not want to bother demonstrating his telephone at the US Centennial Exposition in Philadelphia in 1876; for one thing, he had to grade too many student papers from his speech classes. But Mabel would have none of that. She insisted he go, packed his bags, and packed him off to the train station. When Alexander protested, she turned away. And here again, deafness came to the aid of long-distance hearing. Being unable to read his lips, she literally became deaf to his protests. And so he obediently went.

The rest is history: Bell won the Gold Medal for Electrical Equipment, gained international fame, and founded the Bell Telephone Company with investors, including Mabel's wealthy father, and with much practical business help from the now-unsung Mabel. It was uphill work at first. Typically far-seeing big businessmen of course knew that the telegraph was the only practical means of long distance communication. The Bells had to convince them otherwise.

Bell kept on working with sound. He invented the audiometer, which gave the world the first way to measure hearing ability. He coined a new word, the *decibel* (the *bel* comes from his name), for the measurement of sound levels. He also invented a means of transforming a beam of sunlight into sound: the photophone. "I have heard a ray of sun," he wrote to his father of this successful experiment, which would presage wireless communication and fiber optics.

Bell and Eugenics: Sign Language, Deafness, and Heredity

THE LATE 1800S AND EARLY 1900S WERE BLIGHTED BY VARIOUS eugenics movements advocating the selective breeding of humans, and Bell was an enthusiastic proponent of these dangerous ideas. He hoped for, in his words, "the evolution of a higher and nobler type of man in America, and not deterioration of the nation." He was also worried about "undesirable ethnical elements" entering into the United States, and advocated immigration controls. Unfortunately, he was far from alone with these views, which also entered into his ideas about deafness. In 1884, Bell wrote "Upon the Formation of a Deaf Variety of the Human Race," which advocated the discouragement of deaf-deaf marriages, especially through banning sign language and thus inter-deaf exclusivity. He was a staunch advocate of mainstreaming deaf people into the general population by teaching them to speak and read lips. Of course, sign language advocates vehemently contested Bell on this.

The battle between orality and signing continues in different forms to this day. Signing advocates argue that early acquisition of language skills is essential for cognitive development, and that signing is a natural place to start for a young deaf child—"orality" works well only for those whose hearing is mildly or moderately impaired. Oral advocates argue that sign language reliance cuts deaf people off from regular society. The debates over this can get quite intense.

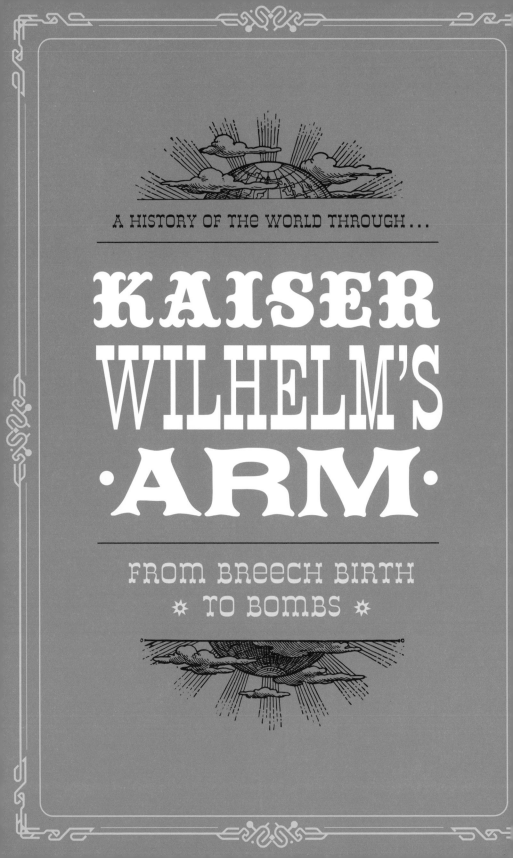

A HISTORY OF THE WORLD THROUGH...

KAISER WILHELM'S ·ARM·

FROM BREECH BIRTH
✳ TO BOMBS ✳

(1859–1941)

C AN WE BLAME WORLD WAR I on one man's arm?
Possibly. Perhaps. Partially.

The arm in question was partially paralyzed as well
as shortened and was the source of much bitterness
to its owner, Kaiser Wilhelm of Germany. To compensate (or over-
compensate, as the case may be), Wilhelm cultivated an uber-manly
image, throwing himself into macho pursuits like horseback riding
and soldiering. Problem was, in spite of his attempts, he wasn't very
good at the latter. The German General Staff derisively said that he
couldn't "lead three soldiers over a gutter." But Wilhem did succeed in
leading, or at least helping to lead, Europe *into* the gutter—World War I.

Wilhelm was born unlucky, a difficult "breech birth," with his bottom instead of his head appearing first. This upside-down entry into
the world pretty much symbolized his future. Up to then, all seemed
propitious, at least in terms of his royal parents' hopes for European
peace and comity that were vested in him. Wilhelm was the heir of
Crown Prince Friedrich Wilhelm (later Friedrich III) and Victoria,
Princess Royal (later Queen) Victoria of Germany, daughter of the
British Queen Victoria. This German-British marriage symbolized what
the two royals wanted: an end to Prussian-style German militarism

and the incarnation of a new British-style liberal Germany. In their son, they had a symbolic joining of two of the most powerful and rising states in Europe. But it didn't work out that way, starting with that backward entrance.

Interestingly, that entrance might not have happened but for Wilhelm's high station in life. According to some, the obstetricians attending his royal mother were forbidden to expose, heaven forbid, her "private parts." They allegedly had to work deep beneath the Princess Royal's voluminous skirts, which naturally made things a tad difficult. Even worse, Queen Victoria of England had insisted that her own British society doctor attend the birthing alongside the German doctor. A problematic decision: Sir James Clark was an elderly specialist—not on obstetrics but on climate and health. (He was also, allegedly, incompetent. Some said that he diagnosed the unmarried Lady Flora Hastings as pregnant when in fact she had a cancerous abdominal tumor.)

Wilhelm was born blue from lack of oxygen, his left arm withered and wrapped around his neck. His shoulder nerves may have been injured during the under-skirt forceps delivery, a so-called brachial plexus injury causing what is known as Erb's palsy, although another recent theory speculates that Wilhelm suffered from intrauterine growth restriction. Doctors had to vigorously rub the newborn future Kaiser to get him breathing, leading some to say that Wilhelm suffered brain damage. Although Wilhelm was intelligent, he had an unusually difficult personality; he was prone to tantrums, nurse-biting, and insisting on his own way. Whatever the cause of his birth difficulties, Wilhelm himself had no doubts. "An English doctor crippled my arm," he said. So began an angry, complex relationship with one thing brachial—and all things English.

Wilhelm's parents had wanted a "perfect" royal son, and they were going to create one, whether he liked it or not. This was a new era for royals. Across Europe, young royals were being educated according to liberal, disciplined, "scientific" middle-class values. The future for European royalty and for Europe seemed bright—as long as the more emotional aspects of nationalism and war could be kept at bay. Scientific medicine was also on the rise, sometimes with wrong turns: Queen Victoria of England had her children assessed by a top phrenologist (a "scientific" head bump and shape reader). He diagnosed her son and future king Edward VII's head as "feeble and abnormal," and not-so-doting mother Queen Victoria concurred, saying he had a "small, empty brain."

In a similar vein, Wilhelm was put on a "scientific" arm-strengthening regimen that would make a saint scream, and that might account for at least some of his warmongering. Doctors prescribed a strong dose of electricity to be passed through toddler Wilhelm's arm daily; they later bound his right "good" arm to his sides to force him to use his withered left, making it difficult to balance and walk and leading to knee dislocations. At four years old, Wilhelm was strapped to a special machine equipped with a metal rod to straighten his back and with a screw to alter his head position. Worse yet, twice a week Wilhelm had to undergo a rather macabre treatment straight out of a grade-B monster movie: placing his arm into the body of a recently killed hare for half an hour to (theoretically) transfer the vital living force of the newly dead animal into him.

None of this worked, nor did his parents' attempt to inculcate "proper" British liberal values. But they kept on trying, maybe too hard, because they were concerned for Germany's future. By Wilhelm's teen years, the great German chancellor Otto von Bismarck had succeeded in uniting most of Germany under Prussia. Much of this had been accomplished militarily, including a quickie 1870 war with France—and now (once he had won) even Bismarck (and Wilhelm's parents) wanted it all to stop.

Wilhelm's parents wanted Wilhelm to do the stopping. If they couldn't get a perfect physical specimen, by Gott, they were going to get the perfect liberal intellectual. Wilhelm was pushed into a liberal educational curriculum that definitely was not liberal in other ways. He had classes 12 hours a day, from 6 a.m. to 6 p.m., six days a week

A Portent of the Future

AFTER THE WAR, DIAGNOSING THE NOW EX-KAISER WILHELM AS insane or as a deranged warmonger became big business among the practitioners of the new science of psychiatry. Doctor and author Ernst Müller did a major study on Wilhelm in 1927 and concluded that he was "high-bred degenerate . . . [with] psychopathy and neurasthenia." Not the kind of leader to win a war. But Müller had a solution to this problem: Germany needed nonaristocratic rulers with ruthless, dictatorial bents, "men with blonde hair, slender heads, blue eyes, of good intellect, of noble sentiment, of lean build, self-confidence and restraint and elegant gait." Sadly, Müller got much of his wish a few years later, although Hitler was definitely not up to this strange Aryan ideal in physiognomy.

under philologist George Ernst Hinzpeter (whom Wilhelm once called a *scheusal*, a monster).

Not only did this not help make Wilhelm into the perfect British-style liberal, but it also didn't make him too thrilled with his parents. Then there was another issue as well, specifically with his mother: Wilhelm apparently had an odd amatory (and somewhat off-putting) fascination with his English mother's hands, and would send her quasi-erotic letters. She would apparently read these letters and coldly respond by correcting Wilhelm's grammar. Presumably, this British stiff upper lip gone too far didn't improve Wilhelm's attitude toward things English.

By the time Wilhelm entered the University of Bonn, he was essentially a full-blown German militarist. He forswore civilian clothes for military uniforms and ultimately owned 120 of them. (He had his clothes made with higher pockets to disguise the shortness of his left arm. Photographs of the militaristic Wilhelm always disguised his left arm; usually his gloved smaller hand was set on a sword. Any photo showing his arm as it really was, Wilhelm had destroyed.) By constant training, he became an able horseman, and through his enthusiastic off-campus military service, he cultivated a new harsh military persona, barking out orders and ideas to everyone around him. But according to his military adjutant, Wilhelm's demeanor was missing one important thing: self-discipline.

Wilhelm succeeded to the throne with few of the royal virtues except a quick intelligence. He kept busy bothering his British cousins—once in London he angrily stated: "You English are mad, mad, mad as March hares." (British Lord Salisbury in return described him in his understated British way as "not quite normal.") Wilhelm began challenging Britain everywhere he could: at sea with a naval buildup, on land with colonies to counter their empire. And he didn't stop there. He developed a strong megalomaniacal streak, even writing: "Deep into the most distant jungles of other parts of the world, everyone should know the voice of the German Kaiser. Nothing should occur on this earth without having first heard him."

People in Europe did hear his voice . . . saying the wrong things. You might say the bitter Wilhelm had an exceptional talent for insulting other European leaders and a knack for finding physical characteristics to insult. He publicly called the short king of Italy "the dwarf," openly made rude remarks about the Bulgarian tsar's nose and spread rumors that he was a hermaphrodite, and called Tsar Nicholas of Russia a "ninny." He was clearly not his parents' avatar for peace.

Worse, in line with the network-crazed spirit of the times, he got Germany deeply embroiled in the tangled web of alliances and counter-alliances that were crisscrossing Europe and finally led to World War I. These alliances theoretically promoted safety: Who would dare go to war with, say, Austria-Hungary if she were allied with more powerful Germany? But we know the answer to that. After the assassination of Archduke Ferdinand of Austria-Hungary, virtually every nation in Europe took up that dare. Like so many bullies, belligerent Wilhelm was reluctant to sign papers declaring war once it seemed inevitable. He was little involved in the actual war beyond bestowing medals (using his good arm), and after defeat, he abdicated and fled to Holland and exile.

Ironically, more effective treatments for the cause of much of Wilhelm's distress and his subsequent political belligerence—the nerve injury that had withered his arm—were refined during the very war he had helped start. Almost 2 percent of injuries during World War I were to peripheral nerves, allowing doctors to experiment successfully with new therapies. A little too late for Kaiser Wilhelm . . . and for the world.

The Kaiser-Einstein Connection

WILHELM HAD AT LEAST ONE POSITIVE ACHIEVEMENT DUE TO his interest in science. Under his reign, the Kaiser-Wilhelm-Institut für Physik was founded. (It later became known as the Max Planck Institute for Physics, or MPP, specializing in high-energy physics and astroparticle physics.) Its first director? An up-and-coming but still relatively unknown young physicist named Albert Einstein.

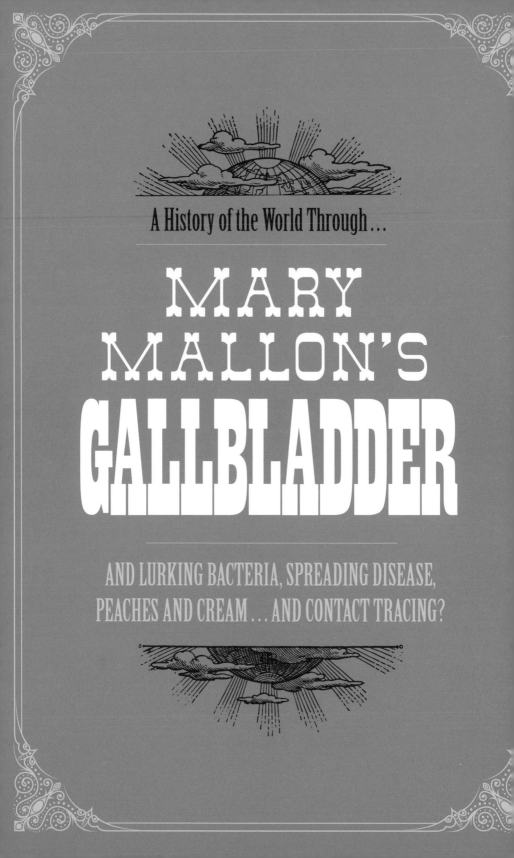

A History of the World Through...

MARY MALLON'S GALLBLADDER

AND LURKING BACTERIA, SPREADING DISEASE, PEACHES AND CREAM... AND CONTACT TRACING?

(1869–1938)

ITY THE POOR GALLBLADDER. It doesn't figure much in history. Why would it? It's basically an innocuous little pouch that lies under the liver and stores greenish bile that helps with fat (lipid) digestion. It's not vital—remove it, and you'll probably be just fine. And in fact, doctors wanted to do just that to the gallbladder belonging to one apparently healthy Mary Mallon, better, if a bit unfairly, known to the world as Typhoid Mary.

Mary Mallon and her notorious gallbladder (in which typhoid germs were furtively lurking) mark an important beginning in modern epidemiology, the science of disease outbreaks. Mallon was the first asymptomatic typhoid carrier—one of the first of any disease, actually—to be scientifically identified and studied. As such, she helped raise questions we still face today: How do we limit contagious diseases, particularly when dealing with asymptomatic carriers? How do we balance individual and group rights? Good questions. And a lot of controversy.

Mallon was born in Ireland in 1869. At age 15 she emigrated to the United States and began working as a domestic, first as a maid and eventually moving into cooking. She became known for her cooking skills and earned the good wage of $45 dollars a month. On August 4, 1906, Mallon found herself on the train to luxurious Oyster Bay on New York's Long Island to cook at the elegant rented summer house of a wealthy banker named Charles Henry Warren. It was there that everything changed.

On August 24, Warren's young daughter fell ill with fever and cramps. The diagnosis was dreaded typhoid fever. (In those pre-antibiotic times, typhoid killed about 10 percent of those it afflicted.) Soon Warren's wife fell ill, then a gardener along with two maids and another Warren daughter, totaling 6 out of 11 people at the house by September 3. Three weeks after the outbreak, Mallon upped and left without giving notice.

Not only the Warren family but also the owner of the rented house became very concerned—typhoid was usually associated with squalid slums and dirt, not with rich people paying top dollar for stays in fancy summer homes. George Soper, a sanitary engineer, was quickly hired to investigate. He found there was nothing the matter with the house or Oyster Bay, but . . . there was something about Mary. She had a suggestive employment history, as in typhoid outbreaks following her from job to job. Just like this one, they hadn't happened among the usual typhoid suspects, but among the very wealthiest New Yorkers.

George Soper had just performed one of the first instances of the important disease control protocol that we now know as "contact tracing." In Soper's words:

It was to be Mary Mallon's fate to clear away much of the mystery which surrounded the transmission of typhoid fever and to call attention to the fact that it was often persons rather than things who offered the proper explanation when the disease occurred.

Back then, the world was in the throes of a revolutionary change in understanding how diseases like typhoid could be controlled. Along with typhoid, cholera, polio, and yellow fever were still periodically sweeping through city populations. To combat this, new sophisticated

What's with That Gallbladder Anyway?

WHY GALLBLADDER BILE, WHICH IS ESSENTIALLY A NATURAL soapy detergent, doesn't kill typhoid bacteria was long perplexing to scientists. But now a study headed by Dr. John Gunn of Ohio State University has an answer.

Once a person is infected, some typhoid germs sometimes escape immune surveillance and often collect on rocky gallstones in the gallbladder, where they form into tough biofilms. Biofilms occur when free-floating microorganisms come in contact with a surface and form a sticky, dense, tough mesh. Once they've stuck onto something and collected and reproduced, they're very tough to kill and uproot—think dental plaque. The immune system essentially gives up trying to completely clear it. Typhoid in the gallbladder is able to remain there, occasionally shedding a few members of the colony into the bile, and so go on to infect others.

public health initiatives began, but the emphasis was still on dirt and hygiene—disposing of garbage, installing better sewer systems, and using water filtration. Cleanups clearly reduced disease; water filtration in particular had drastically reduced typhoid outbreaks. (By 1913, the rate had been cut by more than half, even though it usually took a number of outbreaks to convince penny-pinching city leaders to do something.) But hygiene wasn't enough. Sporadic outbreaks were still occurring, even in "clean" places like Oyster Bay.

Enter an asymptomatic carrier such as Mary Mallon, walking, chatting, cooking, apparently healthy . . . yet shedding toxic germs right and left. This silent-carrier concept wasn't unknown to scientists, but no one was talking much about it, although here too times were changing. Germ theory was already taking over the cruder "dirt concept" of disease control. As for typhoid specifically, scientists had discovered the pesky little microorganism causing the disease, *Salmonella typhi*, back in the 1880s. By now, at least, they could actually identify carriers, even if they displayed no symptoms, by microscopic examination of their feces.

This is exactly what George Soper thought he'd do after he finally tracked Mallon's current whereabouts working as a cook in a fancy Park Avenue home. Not coincidentally, a maid and the owner's daughter had recently become infected. Soper probably wasn't the diplomatic type, although he claimed he was as "diplomatic as possible." But you don't just up and ask a woman you don't know for personal excrement samples. Mallon reacted . . . a little negatively. More precisely, as Soper explained, "she seized a carving fork and advanced in my direction . . . I felt rather lucky to escape."

Soper then enlisted officials from the New York City Health Department and, with the help of five uniformed police officers, finally removed a kicking and screaming Mallon from the house. She kept insisting that she had never had typhoid, but forcible fecal analysis proved otherwise. Mallon was indeed a carrier of the dreaded disease.

Now the second aspect of Mallon's unwilling contribution to the new science of epidemiology: She had already proved that asymptomatic transmission occurred, but a new question arose. What to do with carriers? New York public health authorities had their drastic answer: forcible quarantine. Mallon was put in an "isolation cottage" near Riverside Hospital on an East River island. Although she still denied being a carrier (in fact, according to some accounts, she enlisted a boyfriend's feces as secret replacements to "prove" that she didn't have the bacteria—after all, what are boyfriends for?), she was forced to stay put.

Wasn't this punishing a victim? No, argued New York health officials, we've got to keep the disease under control. But upwards of 3 percent of all recovered typhoid patients were now discovered to be asymptomatic carriers of the disease. Do you quarantine them all forever? No. In 1910, a new health commissioner decided to release Mallon as long as she promised to never cook again. But instead of accepting a city-provided job, she upped and disappeared yet again. Soper and a famous New York physician, Dr. Sara Josephine Baker, finally found her cooking under an alias in the maternity ward at New York's Sloane Maternity Hospital. In the three months that she had been there, 25 people had contracted typhoid; two had died. A study in the *Annals of Gastroenterology* says, ultimately, upwards of 3,000 people may have been infected throughout the city because of her. Mary Mallon was hustled back to her isolation cottage.

Now what to do? A number of solutions had been offered to Mallon, including a touted curative drug, hexamethylenamine (which sounds suspiciously like another touted cure for another disease of 2020 and, like it, didn't work). Next, doctors suggested gallbladder removal. For some reason, the gallbladder is the preferred hiding place for the typhoid bacteria in asymptomatic people. It releases bacteria-laden gallbladder bile into the intestines, the bacteria go from there into feces, then onto hands and from there onto food. (In fact, in the Warren family case, Soper suspected Mallon's trademark dessert, ice cream with cut peaches, was the main culprit. Cooking destroys bacteria, but the peaches weren't cooked and, sad to say, Mallon didn't seem to be washing her hands too well.)

Typhoid, Teeth, and the Fall of Classical Greece

IN 430 BCE, GREEK HISTORIAN THUCYDIDES DESCRIBED A terrible "plague" that swept through Athens, killing a third of its citizens, and ultimately hastening the end of classical Greek civilization. The symptoms as described by Thucydides (who caught it himself) were horrifying, including fevers that were sometimes so intense that the victims preferred to be naked rather than have clothing touch their skin, "unceasing thirst" that couldn't be quenched, violent vomiting, and restlessness so pervasive that many couldn't sleep. Many sufferers died within two weeks, among them, the great statesman Pericles.

But what was this disease? Ebola? Bubonic plague? Historians speculated for centuries, but no one knew until, in 1995, archaeologists excavated a mass burial site in Athens's Kerameikos ancient cemetery. More than 150 dead bodies were unearthed, victims of the plague from more than 2,000 years ago. Scientists extracted teeth from the corpses and further extracted dental pulp from their teeth, and with advanced DNA techniques, finally answered the question: You guessed it, typhoid. Specifically, the DNA sequences of the notorious *Salmonella enterica serovar Typhi* were found, clearly showing that typhoid helped end the Golden Age of Greece.

Mallon vehemently objected to gallbladder removal, a not unreasonable objection given the high mortality rate in those pre-antibiotic days of invasive surgery. She was stuck, then, with the alternative, and was kept in isolation for the rest of her life. She blamed her situation on the rich: "They want to make a showing; they want to get credit for protecting the rich, and I am their victim."

To be fair, they were all facing an intractable problem. Worse, as scientists began studying asymptomatic carriers nationwide, they realized the frightening scope of this: There were probably thousands of asymptomatic typhoid fever carriers nationwide (they estimated 1,300 new silent carriers were created every year) and far worse, *hundreds of thousands* of probable silent carriers of diphtheria and other diseases. They simply couldn't quarantine them all.

With typhoid, thanks to modern vaccines and antibiotics, the problem is mostly over, sadly, long after Mary Mallon's death. But, unfortunately for the rest of us, there are always other, new diseases . . .

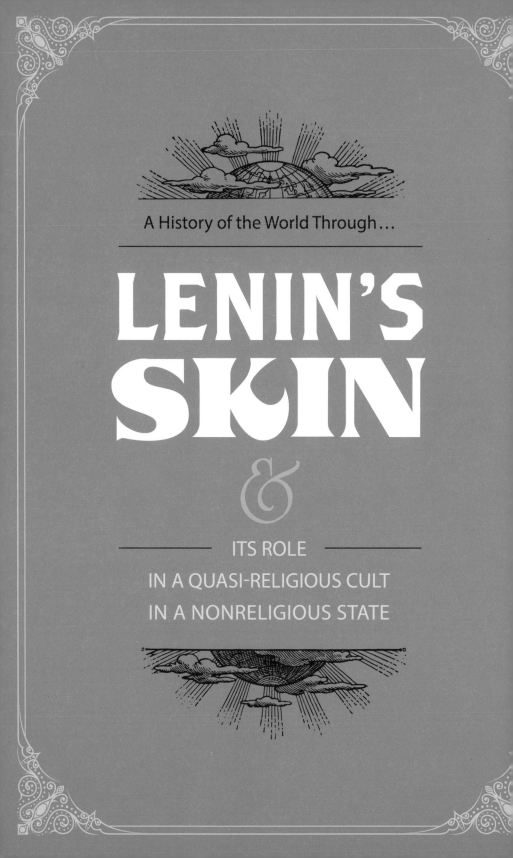

A History of the World Through...

LENIN'S SKIN

&

ITS ROLE
IN A QUASI-RELIGIOUS CULT
IN A NONRELIGIOUS STATE

(1870–1924)

T WAS 1938 DURING THE November holidays, the height of the murderous Great Purge in the communist Soviet Union. Hundreds of thousands of citizens had been sent to their deaths. And Soviet dictator Joseph Stalin and his politburo were paying a visit to Lenin's mausoleum in Red Square.

There the preserved corpse of Vladimir Lenin, founder of the Soviet Union, lay perpetually in state in a crystal sarcophagus. And there the four official mausoleum scientists and workers who maintained the body nervously watched Stalin and his politburo members examining Lenin's pickled skin. Then Premier Vyacheslav Molotov, who had signed more "shooting list" death warrants than even Stalin, muttered the chilling words, "Greatly changed."

Oy blyat! Had they failed? Did Lenin look . . . *dead*?

In Stalinist Russia, failure, even perceived failure, was not acceptable. (As Aleksandr Solzhenitsyn recounted in *The Gulag Archipelago*, people at a speech honoring Stalin were afraid to stop applauding for fear they'd seem anti-Stalinist. They kept nervously clapping until one brave—or exhausted—man finally quit. For this failure of Stalinist zeal, he was arrested for anti-Soviet activities.) And here in the mausoleum,

Stalin's After-Death Beautification Mummification

STALIN'S EMBALMERS FACED A PROBLEM QUITE DIFFERENT FROM Lenin's: They had to make Stalin *less* lifelike. In life, Stalin's face was badly pockmarked and mottled from a serious bout of smallpox. He was short (US president Harry Truman called him a "little squirt"), had a withered left arm, and was rumored to have six toes on each foot. The carefully airbrushed magazine and official photos and portraits always showed a much more handsome and taller Stalin. A member of Stalin's embalming team said in a television interview that his primary task "was to achieve the greatest possible likeness between Stalin's corpse and the photographs and portraits of him so that people would not be shocked." (It didn't really matter in the long run. Stalin lay in state next to Lenin only from 1953 until 1961, when the new Soviet premier Nikita Khrushchev had him buried ignominiously some 300 feet away among the graves of more minor Soviet officials.)

failure could be easily seen. Moldy Marxist skin could mean Siberia or worse.

Keeping Lenin's skin mold-free wasn't easy. The mausoleum crew also had to fight against fungus, "zebraishness" (when the skin developed stripes, spots, or points), and sunken flesh due to dried-out subcutaneous fat. They periodically injected Lenin's face with Vaseline, wax, paraffin, and gelatin to plump it out, shot his muscles up with preservatives, immersed his hands and face in formaldehyde, kept his body in a rubber suit (which was covered by his traditional suit) to keep it constantly bathed in preservatives, and massaged "balsam" (a glycerol and potassium acetate solution) directly into the dead Lenin's skin. This routine had gone on for 14 years, since Lenin died and Stalin first decided it made political sense to attach himself (figuratively) to the dead leader to solidify his position.

Lenin had written a testament shortly before he died that both called for a collective leadership and recommended that Stalin be removed from his position as party general secretary. But it was suppressed by his successors, especially Stalin. Mainly through show trials and executions, Stalin eliminated the others in the post-Lenin collective leadership. He tied himself in the public mind to the revered dead leader by doctoring photographs and commissioning new heroic paintings that emphasized his (false) central role in early Bolshevik communism. And, although the dying Lenin had reportedly said he wanted to be buried alongside his mother, the soon-to-be dictator Stalin would have none of it. A cult of Lenin could help justify his rule. All he needed was a resurrected Lenin to keep the cult going. This led to a rarely asked political question: How do you keep a dead leader's skin fresh?

Soviet doctors and scientists first planned to freeze Lenin's body, but decay set in before a special super-freezer was completed. So they opted for experimental embalming. After Lenin's eyes and most internal organs were removed (his brain was distributed to the Soviet "Brain Institute," created expressly to study his brain for clues to his "extraordinary abilities"), his bodily fluids (about 60 percent of the body by weight) were replaced with preservatives. Only about 23 percent of the original Lenin actually remained, a 23 percent that was carefully protected. At one point, 200 people were working in the mausoleum or adjunct labs on the now almost-holy Marxist body. So valued was Lenin's preserved body (snappily code-named "Object No. 1") that during World War II a special train was sent to besieged Moscow to bring it to a lab set up in central Russia to keep it intact.

This overweening concern for Lenin's dead flesh gives us a surprising snapshot into Soviet life: Probably because the Soviet Union was officially atheist with a great contempt for concepts of the afterworld, the idea of lying lifelike in a permanent state in this present world was immensely important. In the absence of God and traditional rulers, new historical legends had to be nurtured and preserved, figuratively and literally. Lenin's mausoleum was a focal point of adulation with an incorruptible saint of communism eternally lying in state. Over the years, about 24 million people visited, gawked at the body, and mostly marveled at this communist achievement. The Soviets were proud of their expert embalming of their founder and the cult around this one preserved body.

But for all the embalmers' efforts and skill, we are still left with Molotov's displeasure at the state of Lenin's body back in 1938. The much-vaunted Soviet preservation of skin and subcutaneous fat wasn't working as well as was hoped. Lenin's face, lacking the correct amount of subcutaneous fat, kept on sinking; his skin, for all the bleaching and carotene-coloring follow-ups, was often sallow and, worse yet, according to some accounts, chunks of skin and deeper flesh (including part of his foot, fortunately concealed) would occasionally fall off. What to do?

Just after Molotov's visit the scientists came up with a clever nonscientific and relatively non-Soviet answer based on famous (bourgeois) artists. They compared the lighting in Rembrandt's paintings to those of El Greco's, then looked at Lenin's body in the mausoleum and realized that Lenin was illuminated in an unflattering El Greco style. The mausoleum team went wholly Rembrandt, solving the appearance

The Problem with Mummifying Mao

FOR THE PURPOSES OF COMMUNIST EMBALMING, FOUNDING father of the People's Republic of China Mao Zedong died at an inopportune time—in 1976, at the height of Soviet-Chinese tensions. While expert Soviet technicians had been happy to help embalm/mummify leaders of other communist countries, they didn't want to help the Chinese. So the Chinese did it themselves and, by most accounts, they weren't quite up to snuff with their quickly acquired mummification techniques. Mao's embalmed corpse looked . . . wrong. Mao's doctor Li Zhisui reportedly complained that Mao's head "had swelled up like a football" and "the formaldehyde oozed from his pores like perspiration." For about a year, they tried to improve it, but the end result was still only marginally satisfactory. Mao's ears stuck out at odd angles and his skin looked, in the words of a British newspaper, "waxily implausible . . . a Madame Tussaud's reject." Maybe this is why visitors were kept 20 feet away from the sarcophagus and not allowed to linger.

The Not-So-Great
Preservation of
Abraham Lincoln

PUBLIC VIEWING OF EMBALMED LEADERS WAS NOT JUST FOR communists. After his assassination, President Abraham Lincoln's body went on a 180-city, 7-state train tour accompanied by a staff of 300, including an embalmer. At each stop, the corpse was put out for public viewing. But there was no refrigeration in those days. The body began to decay and, despite the best efforts of the embalmer, its features collapsed.

It got especially bad during a 23-hour viewing marathon in New York. The *New York Times* concluded: "To those who had not seen Mr. Lincoln in life, the view may be satisfactory; but to those who were familiar with his features, it is far otherwise. The color is leaden, almost brown; the forehead recedes sharp and clearly marked . . . It will not be possible, despite the effect of the embalming, to continue much longer the exhibition . . ." But the train continued on to Springfield, Illinois, where Lincoln's fast-decaying body was finally interred in a receiving vault at Oak Ridge Cemetery. The Lincoln Memorial with its giant statue of a seated Lincoln serves in some ways as a substitute for the communist-style mausoleum with its embalmed great leader, and has one key advantage: Marble does not rot.

issues with a new glass crystal sarcophagus that not only had better climate control, but also far better specialized lighting. Filters made his skin pinker; shadows that emphasized sunken eyes and cheeks were banished.

Which leads to another point: Was Lenin's body's appearance simply fake to begin with? *Illness, Death, and Embalming of V.I. Lenin: Truth and Myths*, a major book on Lenin's mummification by a prominent Russian surgeon, emphasizes the embalming science behind the expert techniques. But other writers principally outside the former Soviet Union now say that the skin and subcutaneous fat of the leader was more an artificial combination of carefully applied plastic, paraffin, glycerine, and carotene along with heavy makeup, enhanced by that flattering Rembrandt-style lighting. Perfect? Maybe. Natural? Nyet.

Today, Lenin—or what is left of him—still lies preserved in his mausoleum, and opinions on his appearance by modern visitors are quite varied. Some say he appears to be naturally asleep in eternal repose. Others disagree. The more polite among them compare the founder of the Russian communist state to a large wax fruit.

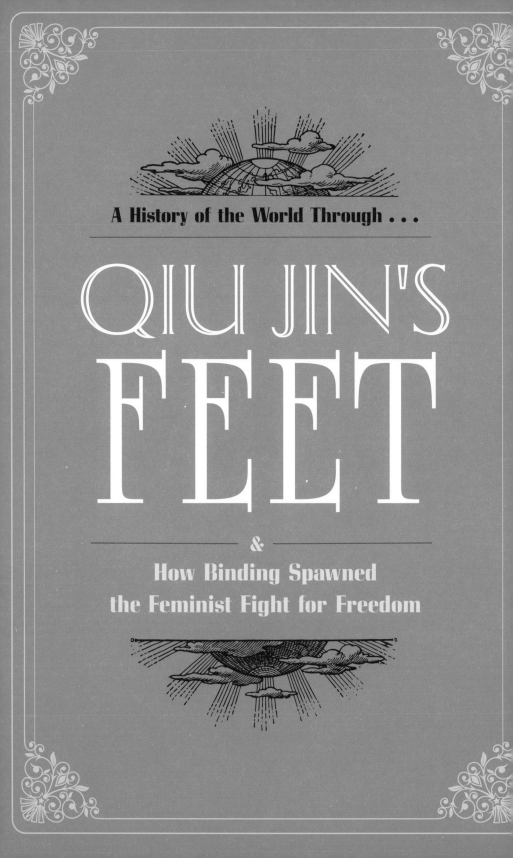

A History of the World Through . . .

QIU JIN'S
FEET

&

How Binding Spawned
the Feminist Fight for Freedom

(1875–1907)

THIS IS A STORY OF how two feet ignited one of the most courageous feminist movements in history. The feet belonged to Chinese feminist revolutionary Qiu Jin, and the movement began because of her shoe size. For upwards of a thousand years in Imperial China, the smaller a woman's foot size, the better. Tiny feet (artificially and painfully downsized feet, ideally less than 4 inches for an adult, which is the average size of a one-year-old's foot) meant status, upward mobility, economic stability, and erotic attractiveness—the last to men, who controlled society. Then women like Qiu Jin started questioning foot-related customs.

But let's start at the beginning. By the time of the Ming dynasty (1368 to 1644), it's estimated that about half of all women had their feet bound. Mothers of daughters weren't content to let nature dictate foot size. They embarked on a time-honored practice of foot-binding to reach that miniscule maximum of 4 inches (although up to 6 inches was common). It wasn't easy.

The excruciating process usually began when a girl was between four and six years old. First, her feet were bathed in hot water, her nails clipped very short, and her feet massaged and oiled. Then all of her toes, except the big toe, were broken and were pushed flat against the sole of the foot; her arch was bent double, sole to heel; and, to keep it all in place, her feet were tightly bound in silk strips. The result was an odd-shaped triangular foot with the big toe protruding. Because the bindings were tight and the toes tightly folded over, the blood supply to the extremities was limited and necrosis, dead tissue, was frequent. Toes would sometimes literally slough off. Every few days, the wrappings were unwound to allow any pus to drain, but the feet frequently got badly infected anyway. Sometimes excess dead skin was cut away; other times it was allowed to rot off. Often the girl was unable to walk for several months, but afterward, to further break the arch, she might be forced to walk (or hobble) long distances. Over time, the bindings got tighter and the feet got smaller. The result was a "lotus foot," also called a "golden lily." Although by all accounts bound feet did not smell floral and fragrant, many men claimed to love the odor and would eagerly drink wine from sweaty lotus shoes. They also found the teetering way in which bound-footed women walked very sensual.

Although painful foot-binding practices had gone on for more than 1,000 years, by the mid-1800s an anti-foot-binding movement was underway. Not only were the Chinese fighting against the tradition, but so too were the European missionaries to China, who began campaigning against the practice. (Ironically, these very same missionaries campaigning against the "barbaric" foot-binding probably saw nothing strange or barbaric in tight corsets that sometimes resulted in the lower rib cage bent inward as a part of Victorian feminine fashion.)

Qiu Jin, born in 1875 to an upper-class Han Chinese family, was among the most strident against foot-binding. She railed against this practice (to which she too was subjected), saying that it caused "untold pain and misery, for which our parents showed no pity." By some accounts, the young Qiu Jin secretly (and later openly) unbound her feet, hoping to undo the painful damage but probably causing her yet more pain. But in more ways than one, she was determined to break the old ties that she felt were quite literally binding China. Paradoxically, China at the time was ruled by the Manchu Qing dynasty, which hailed from Manchuria in the north. They were against foot-binding for themselves; their women wore platform shoes instead, giving them a similar delicate walking style without the anguish. But the foot-binding tradition was too strong among the Han Chinese, and it persisted, particularly as a way of distinguishing the Han Chinese from their "barbarian" Manchu rulers.

As Qiu Jin grew up, she became quite accomplished as a poet (one of her poems began, "Don't tell me women / are not the stuff of heroes") but was still also a somewhat traditional and dutiful daughter. At 18 she entered into a traditional arranged marriage with a wealthy merchant who turned out to be a hard-drinking, prostitute-loving, sycophantic and traditionalist social climber. After enduring a few years of marital hell, Qiu Jin chucked it all, fled to Japan, and joined a revolutionary group in a quest to overthrow the corrupt Manchu Qing regime. One of her most important efforts was writing an eloquent polemic, "A Respectful Proclamation to China's 200 Million Women Comrades," which denounced foot-binding and the subjugation of women. She tied much of women's economic and political backwardness to the mobility problem with bound feet: "Being thus necessarily dependent on external aid, we find ourselves, after marriage, subjected to the domination of men, just as though we were their household slaves."

While many women like Qiu Jin were the fiercest opponents of foot-binding, other women were also some of its strongest defenders. Oddly, foot-binding was a great equalizer. A poor girl with tiny bound feet could marry a wealthy man and advance her social status. A woman seen as unattractive in other ways could, with smaller-than-average feet, become highly sexually attractive. Family status was enhanced by having daughters with bound feet—showing that the family could afford women who were relatively economically unproductive and house-bound. And recent research has discovered a foot-binding paradox: In some areas of rural China, there was actually a counter-economic advantage to girls with bound feet. They couldn't wander far, so would stay home doing necessary and dull textile and clothing handiwork. In effect, they were foot-bound economic prisoners.

By the early 1900s, Qiu Jin was not only a feminist but also a revolutionary. With fellow poet Xu Zihua, she started the feminist newspaper *Chinese Women's News,* which encouraged women to become independent, work for themselves, cease being dependent on men, and, of course, eschew foot-binding. She angrily denounced men who were charmed by the "captivating little steps" women were forced to make on their feet. As is often the case with truthful reforming journalism, the paper was shut down after only two issues. Undeterred, Qiu Jin went on to become head of a girls' sports school (which was apparently really designed to train revolutionaries). She seemed to have a bright—and revolutionary—future ahead of her, until she was apparently betrayed by one of her compatriots, arrested, tortured, and then beheaded in 1907.

But her legacy lived on. She was—and is—a national hero in China today. Foot-binding was outlawed in 1912, a few years after her execution, and more importantly, it lost prestige as a custom. Foot-binding in some areas lingered on, particularly in remote rural provinces even after the communist revolution in 1946, but today, only a few very elderly women have feet that were bound. And the last lotus shoe manufacturer (making doll-like shoes for tiny lotus feet was a specialized business—after all, not many regular shoemakers stock size 1 adult shoes) finally closed in 1997.

The Beginnings of Binding

BINDING MAY HAVE BEGUN BECAUSE OF THOSE SEEMINGLY MOST benign of literary types—poets. (Maybe there was a reason Plato banned them from his ideal republic.) Professor Dorothy Ko of Barnard College thinks foot-binding may have begun during the Tang dynasty, which ruled China from 618 to 907. She explains: "One may say the ideal of dainty feet was concocted by male poets in the Tang dynasty, and taken up as an actual practice by women in elite families by the 12th and 13th centuries. The missing step—one that I can only conjecture—is that when the Tang court fell in 907, palace dancers were dispersed to courtesan houses in the south, introducing a mild form of binding—like a ballerina's pointes."

From there the foot-binding regimen got harsher and more widespread. Some estimate that by the 1800s between 40 and 80 percent of Han Chinese women had bound feet—and among the wealthy elite the percentage probably hit near 100 percent. The ideal was a 3-inch foot, a golden lotus. Four inches—a silver lotus—was okay, but no one wanted more. A woman with "huge" 5- or 6-inch "iron lotus" feet (well below a size 4 adult woman's shoe) faced problems in the marriage market.

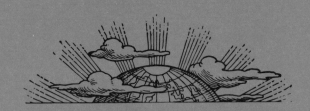

A HISTORY OF THE WORLD THROUGH...

EINSTEIN'S

BRAIN

AND A TALE OF CIDER BOXES, MASON JARS,

AND, POSSIBLY,

THE BIOLOGICAL BASIS OF GENIUS

(1879–1955)

"Einstein was a world-famous genius and people I knew used to remark, 'You spend a good deal of time with Einstein. He has a perfect brain, doesn't he?'"

—physicist Eugene Wigner

WHEN ALBERT EINSTEIN WAS ALIVE, even people who didn't think his brain was exactly perfect suspected it was *different*, better in some way than your average didn't-come-up-with-relativity-and-never-could kind of brain. But different how? No one knew. After all, it's a tad difficult to analyze a brain in its original container. Then, on April 18, 1955, at age 76, Einstein died—and an intrepid pathologist seized the opportunity to seize the brain.

Einstein had been in the Princeton Hospital suffering from intense abdominal pain; on the table beside his hospital bed were scribbled notes, thoughts on how to make the theory of relativity jibe with quantum mechanics, part of his work on the so-called theory of everything. But at 1:15 a.m., his body—and his still-active brain—finally stopped. According to the nurse on duty, Einstein took several deep breaths, muttered a few words in German, and died.

Princeton hastily called a press conference announcing that the genius was dead, while the body was whisked to the lab for an autopsy. Pathologist Thomas Harvey performed his straightforward task to officially determine the cause of death. (As suspected, a long-standing aortic aneurysm had ruptured.) But then, instead of sending the body back to the family for cremation as planned, he added an extra step: He removed Einstein's brain. (In the spirit of collegial sharing, he also removed Einstein's eyes for Einstein's oph-thalmologist.) Only then was the body sent back to the family, who had it immediately cremated later that day and the ashes scattered without any publicity per Einstein's wishes.

The family had no idea that they had received only a partial Ein-stein until the next day when they got their *New York Times*. There, in a front page article with the compelling headline "Key Sought in Einstein Brain," they read that "the brain that worked out the the-ory of relativity" had been removed "for scientific study." Confused, Einstein's son Hans Albert and his executor Dr. Otto Nathan went to confront the improvising pathologist who admitted that, when faced with this unique chance to harvest the brain of such a genius, he took matters—and the brain—into his own hands.

Harvey was following in the footsteps of scientists (and "scien-tists," for that matter) before him who were gripped by a sort of, well, brain fever—the pressing urge to study brains firsthand, particularly those of the famous or, in some cases, infamous. This zeal for gray matter began millennia ago in the 5th century BCE when Alcmaeon of Croton came up with the idea that the brain was the seat of our con-sciousness. From that point on, brains were a focal point of research, and dissection of brains became a crucial method of study. But fads come and go, and for a while cutting up human bodies in the name of science (versus acceptable dismembering in the name of war and nationalism, of course) was frowned upon. It wasn't until the 13th century that scientific human dissection began in earnest again, and it wasn't until 300 years after that, that brain research made a huge leap with the development of soft tissue preservation methods that allowed researchers to collect organs for study.

Brain collections really hit their stride in the 1800s. All things brain-related were hot—from phrenology, the "scientific" study of bumps on the head, to atavistic classification, linking head shape and size to criminal tendencies, to, well, just brains. The study of brains was in high gear. Scientists collected them, analyzed them, and compared them. In the late 1800s, one researcher, Burt Green Wilder, divided his brain collection into two main sections: "Brains of Educated and Orderly Persons," that is, those belonging to (white male only) scholars and notable luminaries, and "Brains of Unknown, Insane, or Criminal Persons," which included not only the mentally ill and criminals, but also minorities and women. Brain societies or "clubs" formed—scientists and interested laypeople banded together to discuss brains, to promise their own brains (post-death, of course) to the club for eventual study, and to convince the famous to similarly bequeath their brains. At the bottom of all this brain fascination was the belief that surely there was some difference in the brains of the best and the brightest, something that distinguished them from other brains . . . which brings us back to Harvey and Einstein's brain.

The family owed the brain to science, Harvey argued. Although no agreement was formally put into writing (one of the reasons Harvey was long dogged by stories about how he "stole" the brain), the family and the executor ultimately accepted Harvey's action and agreed that the brain could be studied. But there was one major stipulation: In keeping with Einstein's desire to keep as low a profile as possible, the family wanted the whole process kept quiet; no public relations blitzes.

They got their wish. Nothing Einstein-brain-related appeared in the media; no studies were published. The brain dropped out of sight and mind until 25 years later, when an intrepid reporter for *New Jersey Monthly* magazine went on a "find Einstein's brain" quest. Reporter Steven Levy ultimately tracked Harvey down in Wichita, Kansas. There, Harvey showed Levy two jars in a cardboard cider carton filled with

CONTINUED >

Bring Me the Brain of Benito Mussolini

ITALIAN DICTATOR BENITO MUSSOLINI'S BRAIN WAS YET ANOTHER noteworthy brain removed from its original casing. After the fascist leader had been killed by a communist firing squad in 1945, doctors at Milan's Institute of Legal Medicine did an autopsy and removed his rather damaged brain. (A firing squad can have that effect, as can being hung, head-down, for the public to kick, punch, and otherwise abuse.) The US government asked for a sample, wanting to check out the theory that Mussolini's bizarre behavior was the result of general paresis, an offshoot of untreated syphilis (and probably also wanting it as a trophy). So test tubes of brain matter winged their way to Washington, DC, where doctors studied it, concluded that the brain was completely normal, and put the brain samples in storage, one vial at St. Elizabeth's Hospital, the other at the Armed Forces Institute of Pathology. And there they stayed, forgotten by most—but not by

Mussolini's widow. She got her husband's missing body back in 1957 and, wanting to bury him in as much of a complete state as possible, wanted his brain back. She kept writing the US ambassador to Italy with this request, and finally, in 1966, government officials agreed. Except they could find only half of the brain samples. The one at St. Elizabeth's Hospital was missing. But half a brain was, apparently, better than none. The brain sample was sent to Florence in a plain white envelope (mis)labeled "Mussolinni" and, presumably, joined the body in the crypt. What of the missing brain? No one is sure what happened to it. But in 2009, a listing appeared on eBay—an anonymous seller was offering, for a relatively inexpensive €15,000, bits of Mussolini's brain. (Upon being notified by Mussolini's granddaughter, eBay removed the listing.)

crumpled newspaper tucked behind a beer cooler: "Floating inside [a mason] jar . . . were several pieces of matter. A conch shell-shaped sized chunk of grayish, lined substance, the apparent consistency of sponge . . ."

And in a second larger jar, with its top held on by masking tape, were "dozens of rectangular translucent blocks, the size of Goldenberg's Peanut Chews, each with a little sticker reading CEREBRAL CORTEX . . . Encased in every block was a shriveled blob of gray matter."

Meet Einstein's brain.

Harvey explained that he had cut the brain into 240 samples and had made microscopic slides that he sent to prominent neuropathologists. While he had kept the rest of the brain for his own research (which he never did), Harvey had hoped and expected that other more prominent scientists would be eager to study Einstein's brain. But many weren't very interested, not in the least because there was some controversy about the ethics of the whole brain-removal incident. It wasn't until decades later than studies were finally published.

So did studying Einstein's brain shed light on the genesis of genius? It's difficult to be certain, but researchers think they've made some notable findings. A 1999 study done by researchers at McMaster University found that Einstein's brain was actually smaller than average, but certain parts of his brain, like his parietal lobes, were larger than average and more highly developed. More than a decade later researchers at the University of California at Berkeley found that he also had a higher ratio of glial cells to neurons and a higher number of connections between all those glial cells. (In layperson's terms, this means that Einstein may have had enhanced cognitive ability, that he could make creative leaps more easily than most.) But this was—and is—conjecture. We're still only beginning to understand how brain structure contributes to intelligence.

Harvey never knew his "theft" of the brain was to some degree vindicated. He died at age 84, years before the more interesting studies had been issued, breathing his last in the hospital where Einstein had died and where he had initially taken the brain.

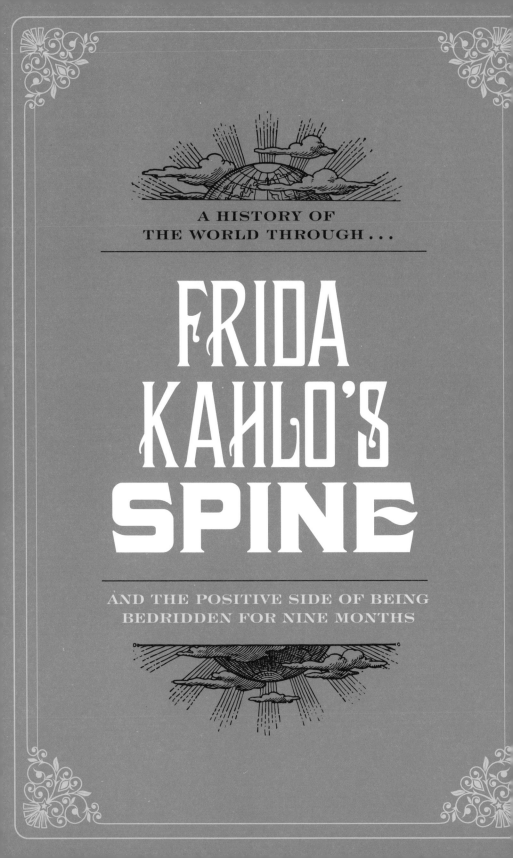

**A HISTORY OF
THE WORLD THROUGH...**

FRIDA KAHLO'S SPINE

AND THE POSITIVE SIDE OF BEING
BEDRIDDEN FOR NINE MONTHS

(1907–1954)

HEN IT COMES TO HISTORICAL body
parts and artist Frida Kahlo, most people would
immediately think "unibrow." It makes sense;
it's featured prominently in her self-portraits
and dominates her face in photographs. (Kahlo actually would enhance
her brow with eyeliner pencil to make it more prominent.) It's her
iconic symbol, her trademark, visual shorthand for the artist herself.

Case in point: When a Frida Kahlo Barbie was released in 2018, there
was intense backlash, not because a feminist painter had morphed
into a preposterously proportioned doll but because the doll's unibrow
wasn't assertive enough, wasn't "Frida" enough. But the much-vaunted
unibrow probably would have remained unknown unplucked facial
hair if it weren't for the body part that really made Kahlo the icon she
is today: her spine.

The then-18-year-old Kahlo and her boyfriend, Alex Gomez Arias,
were traveling on a bus when it collided with a trolley. As Arias later
said, the force of the impact caused the bus to "burst into a thousand pieces." One of those pieces—a steel handrail—impaled Kahlo,
spearing her through the hip and fracturing her spine in three places.

Arias described a scene as fraught with magical realism as one of her paintings:

> Something strange had happened. Frida was totally nude. The collision had unfastened her clothes. Someone in the bus, probably a house painter, had been carrying a packet of powdered gold. This package broke, and the gold fell all over the bleeding body of Frida. When people saw her, they cried, "La bailarina, la bailarina!" With the gold on her red, bloody body, they thought she was a dancer.

It might have looked surreal, but the outcome was harsh reality: To treat her spine, Kahlo was encased in a plaster body cast, unable to walk or stand for three months. (She had dealt with something similar as a child of six when she contracted polio and was bedridden for nine months.) Faced with the prospect of crushing boredom and needing something to distract her from her pain, she had a special easel made as well as a support for her to lean against so she could paint while lying in bed. She also had a mirror positioned above her bed so she could always have a subject to paint, a subject that became a hallmark of her work—herself. And so Frida Kahlo, artist and original selfie superstar, was born.

Beginning in that hospital bed, she captured herself over and over again on canvas in pre-smartphone (and much more complicated) versions of selfies. By the time she died in 1954, she had painted 55 self-portraits, most of them not just straightforward depictions of herself, but with surrealistic overlays and accoutrements in a kind of intellectual, metaphorical, and sophisticated version of a Snapchat filter.

Actually, Kahlo's self-portraits represent only 34 percent of her paintings. She painted 88 others that didn't star her, but rather highlighted aspects of Mexico, a major love of her life. Kahlo was so attached to Mexican independence that she even changed the date of her birth. Born in 1907, she told people that she had been born in 1910. This was no vanity whim to make her appear three years younger, but instead

Historical Unibrows

KAHLO WASN'T THE ONLY PERSON WHO ENHANCED HER UNIBROW for dramatic effect. In ancient Greece, women would use antimony, a dark gray metalloid, to make their eyebrows look thicker and joined together in what they called a synophrys (from *syn*, "together," and *phrys*, eyebrow). A unibrow was considered a sign not only of beauty but also of intelligence and was extolled by writers like the ancient Roman Petronius, who once wrote that his ideal woman would have eyebrows that "almost met again close beside her eyes." (Yes, almost. It seems that modified almost-unibrows were the most fashionable. As writer Anacreon described his mistress, she had brows that "neither join nor sever.") From that point on, unibrows went in and out of fashion in different times and countries. They were the rage in India (pictures of goddesses and queens typically sport dark unibrows) and were very out in ancient Persia, where threading to tame eyebrows and other facial hair was the vogue, but came back in during the Qajar dynasty in the 1700s. In Europe, unibrows were pretty much out, because foreheads, particularly high foreheads, were The Thing. Women plucked their eyebrows to be pencil-thin to create that ultimate in medieval beauty: a high-domed forehead.

was done to emphasize her ties to Mexico: 1910 was the 100th anniversary of Mexican independence.

But it's the self-portraits that most people think of and have seen, at least as reproductions. They're now ubiquitous, with mugs and T-shirts, Band-Aids and kitchen towels all featuring some version of staring-straight-at-you-from-beneath-heavy-unibrow Frida. This Frida-mania is a distinctly modern phenomenon, something that started in the late 1970s when her work was rediscovered. Before that, she was known outside of art circles not as a famous artist but as a *Mrs.* Famous Artist, specifically Mrs. Diego Rivera. In fact, the headline of her *New York Times* obituary reads, "Frida Kahlo, Artist, Diego Rivera's Wife." (They did manage to work in a few lines about her art.)

Husband Diego Rivera was a leading man in Mexican art, famed for his politically infused paintings, particularly his massive frescoes. His and Frida's marriage was similarly political (both were communist activists and Mexican nationalists). It was also famously complicated. Rivera was known as a womanizer and began having affairs shortly after they married. Kahlo retaliated by having affairs of her own, with both men and women (some of whom reportedly also slept with Rivera) and it became almost like a series of one-upmanship. After Rivera had an affair with Kahlo's sister, Kahlo had an affair with Rivera's idol, Soviet revolutionary Leon Trotsky. The infidelities finally led them to divorce in 1939, ten years after they had married, but they remarried only a year later. Apparently they couldn't stay apart in spite of all the drama. Kahlo once said, "There have been two great accidents in my life. One was the trolley, and the other was Diego. Diego was by far the worst." (Rivera intriguingly called her "the great fact of his life.")

While Rivera dominated her life in one way, Kahlo's spine did in another. It was a chronic problem for the rest of her life. Not only did she have surgery on it more than once (she had 30 surgeries in total, some on her legs, others on her spine), but she also wore plaster corsets on and off throughout her life. Just as she turned her bedridden recuperation into an opportunity to create art, she did the same with the corsets. She painted on them and covered them in printed fabrics like a molded collage. Her corsets were mélanges of trolleys, monkeys, birds, and tropical plants. When she wore them, she was a walking canvas, a human painting, a 3-D version of one of her self-portraits. And her self-portraits often captured her health, among them *The Broken Column*, an overt reference to her spine.

In April 1953, Kahlo's dream finally came true. She was having her first solo exhibition in Mexico. No one thought she would be at the opening reception to revel in her triumph; she was on doctor-ordered bed rest. And speaking of beds: What was that four-poster bed doing in the middle of the gallery, attendees wondered. They didn't wonder long. An ambulance pulled up to the gallery. Kahlo was wheeled out on a stretcher. The crowd inside parted as she was placed on the bed that was waiting for her. And there she held court, the unconquerable Frida, greeting her guests and patrons.

The "Little Woman" Artist

FOR MUCH OF THEIR MARRIED LIFE, RIVERA GOT THE BULK OF the attention and fame, even as Kahlo continued painting. When in 1932 they went to Detroit, where Rivera had been commissioned to paint a series of frescoes on industry, reporters asked the 25-year-old Kahlo if she too was a painter. "The greatest in the world," she told them. The response and her attitude got the limelight on her ... kind of. A reporter decided to do an article on her, not Rivera, for the *Detroit News*. But the article praised her rather backhandedly, commenting that she was "a painter in her own right" whose work was "by no means a joke." The title of the piece? "Wife of the Master Mural Painter Gleefully Dabbles in Works of Art." A few years later, in 1939, a *Life* magazine article about Rivera's mural at San Francisco's City College that included Kahlo as a subject refers to her simply as Mrs. Diego Rivera.

It wasn't an uncommon bias in the art world. Other artists who were wives of fellow artists also fell into the Mrs. Artist trap, even years later. Realist Edward Hopper's wife, Josephine Hopper, was a successful painter who showed with luminaries like Picasso and Modigliani. But once his career took off, hers faded. She is now better known as the woman in all of her husband's paintings. In the 1940s and '50s, abstract expressionist Lee Krasner was overshadowed by spouse Jackson Pollack. When people think of de Kooning, they typically think of Willem, not Elaine. Even though she had success, particularly as a portraitist (she was chosen by JFK to do his portrait), she isn't the default de Kooning.

In the case of Kahlo, though, time changed the power balance. It's she who now speaks more to the masses. There has been no attempt to set a world record for Diego Rivera look-alikes; there are no Diego Rivera sneakers, watches, or flowerpots, no films named *Diego*. In fact, the curator of a recent exhibit of the couple's work commented that it was Rivera who needed an introduction to the viewing public, not Kahlo.

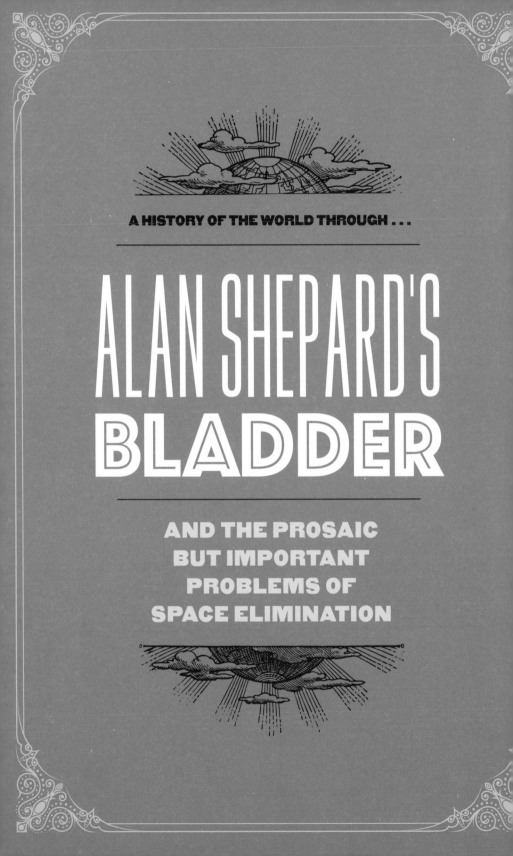

A HISTORY OF THE WORLD THROUGH...

ALAN SHEPARD'S
BLADDER

AND THE PROSAIC
BUT IMPORTANT
PROBLEMS OF
SPACE ELIMINATION

(1923–1998)

ACK IN 1961, DR. FREEMAN H. QUIMBY of the Office of Life Science Programs at NASA had categorically stated that "the first space man is not expected to have 'to go.'" After all, the flight of Alan Shepard, the first US astronaut to go into space, was going to last only 15 minutes. So not a problem, right?

Not quite. NASA didn't take into account the time *before* takeoff. A launchpad delay had kept Shepard in his space suit seated in the capsule cone for eight long hours and his bladder was now very full. Dr. Quimby's confident pronouncements notwithstanding, he really, really *did* have "to go." And there we have it, that perpetual problem not just of bodily waste elimination but of being human in general, with the requisite all-too-human body that we all too often ignore in our planning, as in the case of that patiently waiting bladder of Alan Shepard in his space capsule.

When Shepard could hold it no longer, he informed NASA of his impending liquidity problem. What to do? Shepard was told in no uncertain terms to stay put and stay strapped in his capsule seat. He then asked Mission Control for permission to go in his suit, which was duly granted. It really wasn't that bad, Shepard remembered: "Of course with a cotton undergarment, which we had on, it soaked up immediately. I was totally dry by the time we launched." But it was a problem scientifically. By peeing, Shepard shorted out the electrical system of the medical sensors designed to track his physiological reactions during the flight.

Shepard's bladder raised an important question relating to man's (and later woman's) sojourns in space. Just how do we take care of the inevitable bodily functions, including that of elimination, in the heavens above? It's not sexy stuff, but it's one of the major challenges we face as we leave our familiar terrestrial home—which, of course, is well equipped with useful accoutrements such as urinals and toilets.

But first things first. After Shepard's problematic (but otherwise successful) flight, NASA got to work on the pressing problem of peeing in space suits. And here we get into a problem when trying to delve into this backwater, if you will, of history. Peeing in your clothes on the job has not been a particularly potent human problem historically, except perhaps for one particular moment in history—that of knights in shining armor, specifically plate armor of the 15th and early 16th centuries.

Knights suited for battle were protected from the slings and arrows of their enemies. But even in the heat of combat, sometimes nature called. What then could Sir Knight do, canned, as it were, in his armor? On the plus side of things, a suit of plate armor wasn't as all-encompassing as a space suit. Because a knight needed to be able to ride effectively and comfortably, his armor wasn't an all-in-one jumpsuit made of metal. His armor was actually in pieces, typically consisting of a helmet, breastplate, gauntlets, steel legplates, padded breeches, sometimes with chainmail underpants, and to protect his nether regions, mail skirts (often articulated) consisting of a fauld covering the front hips and a culet covering the backside. As such, he (theoretically) could relieve himself without a can opener.

On the negative side, it wasn't easy. Metal gauntlets, even when articulated, could allow a knight to lift his skirts, but it was clumsy. (Pity the poor arming squire who had the unenviable job of either helping the knight relieve himself by lifting the faulds or the culets, depending on what side needed to be relieved, or, less pleasantly, cleaning the armor—usually with sand, vinegar, and urine, to save water—if said knight was unable to take a bathroom break from battle.) More importantly, as noted even by modern reenactors, being inside a suit of armor is *hot*; much of the potential pee is eliminated via sweat. Yet if you simply had to go, you'd do what Alan Shepard did, and as in his case, much would have been absorbed, this time by the heavy padding worn underneath the armor. But much of this is conjecture. There's little written about the mechanics of bodily waste elimination; knightly chroniclers preferred to focus on more chivalric deeds.

And back in the mid-20th century, there is a similar gap in the documentary evidence of space-age elimination. (Some research that was published is excessively euphemistic—a contemporary British research paper on the subject, for example, refers to wrapping an impervious sheathing around a man's "urinary duct." Modern science, like the rest of us, calls it a penis.) We do know, though, that for the flight that followed Shepard's, astronaut Gus Grissom was given a space suit with a urine reservoir that worked fairly well, although it was a bit tight; and some astronauts complained of leakage and skin irritation. (Grissom was also denied coffee on the morning of the flight to reduce his urinary output.) By the time John Glenn went into orbit, urine collection was getting better, using modified condoms—"roll-on cuffs"—as an external catheter. Engineers tested a number of store-bought condom brands until they found one that wouldn't leak, and then got the manufacturer to make an even stronger one. Unlike regular condoms, of course, these had holes at the end that stretched around a storage bag held in place by the tight-fitting space suit. According to some sources, the NASA condom sheaths came in three sizes, small, medium, and large. Predictably, all the astronauts asked for large sheaths, so NASA added a twist, sizing the sheaths as "Extra-large," "Immense," and "Unbelievable."

But it all didn't go as simply as expected. After Glenn's trip, and particularly in subsequent space flights, scientists began to learn that

The First Human
to Pee on the Moon

BUZZ ALDRIN, THE SECOND HUMAN TO REACH THE MOON, can proudly boast (as he indeed has) that he was the first to pee there. This dubious honor wasn't planned. According to Teasel Muir-Harmony, curator of the National Air and Space Museum's Space History Department, "When he [Aldren] landed the lunar module, he landed so softly that the legs, which were designed to compress, didn't." So instead of taking a small step from the module to the moon's surface, Aldrin was forced to take a leap—and the jolt when he landed on the moon caused his urine collection device to break. "So instead of going where it was supposed to, the liquid ended up collecting in one of his boots. When he walked around the lunar surface he was kind of sloshing around."

the human body—and the mind controlling it—must make special adaptations. For example, Glenn apparently did not urinate early enough on his first flight. Usually the human bladder begins to feel the urge when about one-third full; at about two-thirds it really feels it's time; beyond that begins pain and acute discomfort. Apparently, astronaut Glenn hadn't felt that urge even though his bladder was more than full. The problem was gravity: In zero gravity, urine doesn't collect at the bottom of the bladder, so the urge to go isn't as strong. But if urine is held inside too long past its due date, the pressure on bladder sphincter valves can damage them, causing permanent incontinence.

And what about female astronauts? First NASA had to get over its misogyny. In 1962, NASA had stated authoritatively that due to female "physical characteristics," women didn't belong in space, ignoring a test that showed that women actually had a 68 percent success rate in fitness testing for space versus 56 percent for men. And a 1964 report notes "intricacies of matching a temperamental psychophysiologic human and the complicated machine." (You know . . . PMS.) Once women were on board, a new urine-collection scheme had to be developed. The Maximum Absorbency Garment, essentially a space-age diaper, was born, and it was soon adopted by men as well.

As for number two, it predictably gets a bit messier. NASA's "fecal containment system" wasn't always up to its job or its fancy name. It was an "extremely basic system" by NASA's own admission, essentially a plastic bag taped onto the buttocks that astronauts could reach by opening the convenient flap in the back of their space suit. (There was also a special onboard compartment for toilet paper.) In an unexpected twist—and to many an "I don't want to think about this" part, so readers are duly warned—due to the absence of gravity, bowel movements had to be manually helped along, and often had to be chopped or pinched off by hand. The bags were equipped with side "finger coverings" to allow astronauts to do so without getting their hands dirty. The whole defecation operation took about 45 minutes. According to astronaut Peggy Whitson (who spent 665 days in space), it was her least favorite part of working in zero gravity. Not surprisingly,

| A Period in Space

IN 1983, JUST BEFORE ASTRONAUT SALLY RIDE'S FIRST SPACE-
flight, NASA engineers (presumably all male) confronted her with a
basic question: How many tampons might she need for her one-week
mission? "Is 100 the right number?" they innocently asked. "No," she
said calmly, "that would not be the right number." (N.b.: The average
number used per period is 20.)

The number of tampons was a small matter. As with the earlier
unexpected problem of zero-gravity urine, (male) scientists were
concerned that menstrual blood might float upward into the abdomen
and create problems even while female astronauts were confident
retrograde bleeding wouldn't occur. They were right. Unlike many
other physiological functions, it appears that menstruation isn't
affected at all by spaceflight—it's essentially the same as on the
ground. Of course, even on the ground it can be problematic, so most
women opt for birth control pills or an IUD.

things could get quite smelly in the cramped quarters. And sometimes there was leakage, leading to these immortal lines from the *Apollo 10* astronaut Tom Stafford in space in 1969: "Get me a napkin quick. There's a turd floating through the air."

Quickly moving away from bowels, a final word: While humankind may indeed reach for the stars (or, in Alan Shepard's case, a little lower, into the ionosphere), we still can't get away from it—we've got bodies that have their own agendas and, well, *needs*. Even humble bladders.

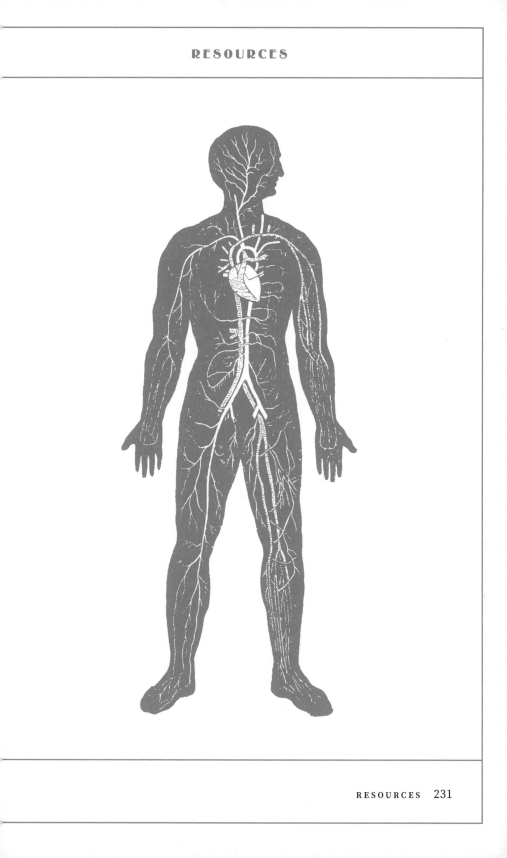

PALEOLITHIC PYRENEAN WOMAN'S HANDS

Basedow, H. *Knights of the Boomerang*. Sydney, Australia: The Endeavour Press, 1935.

Dobrez, P. "Hand Traces: Technical Aspects of Positive and Negative Hand-Marking in Rock Art." *Arts* 3, no. 4 (2014): 367–393. https://doi.org/10.3390/arts3040367.

Groenen, M. "Les représentations de mains négatives dans les grottes de Gargas et de Tibiran (Hautes-Pyrénées). Approche méthodologique." *Bulletin de la Société Royale Belge d'Anthropologie et de Préhistoire* 99 (1988): 81–113.

Gross, C. G., C. E. Rocha-Miranda, and D. B. Bender. "Visual Properties of Neurons in Inferotemporal Cortex of the Macaque." *Journal of Neurophysiology* 35 (1972): 96–111.

Leroi-Gourhan, A. *The Art of Prehistoric Man in Western Europe*. London: Thames & Hudson, 1967.

Petrides, M., and D. N. Pandya. "Distinct Parietal and Temporal Pathways to the Homologues of Broca's Area in the Monkey." *PLoS Biology* 7, no. 8 (2009): e1000170. https://doi.org/10.1371/journal.pbio.1000170.

Romano, M., et al. "A Multidisciplinary Approach to a Unique Paleolithic Human Ichnological Record from Italy (Bàsura Cave)." *eLife Sciences* 8 (2019): e45204. doi: 10.7554/eLife.45204.

"Science Notes: Paleolithic Cave Art and Uranium-Thorium Dating." *Current Archeology*, April 24, 2018.

https://archaeology.co.uk/articles/sciencenotes/science-notes-palaeolithic-cave-art-and-uranium-thorium-dating.htm.

QUEEN HATSHEPSUT'S BEARD

Cooney, K. *The Woman Who Would be King*. New York: Oneworld Publications, 2015.

Izadi, E. "A New Discovery Sheds Light on Ancient Egypt's Most Successful Female Pharaoh." *Washington Post*, April 23, 2016.

https://www.washingtonpost.com/news/worldviews/wp/2016/04/23/a-new-discovery-sheds-light-on-ancient-egypts-most-successful-female-pharaoh/?noredirect=on&utm_term=.953365a29387.

Mertz, B. *Temples, Tombs and Hieroglyphs: The Story of Egyptology*. New York: Harper Collins, 2007.

Robins, G. "The Names of Hatshepsut as King." *The Journal of Egyptian Archaeology* 85 (1999): 103–112.

Tyldesley, J. *Hatchepsut: The Female Pharaoh*. London: Penguin Books Ltd., 1998.

Wilford, J. N. "Tooth May Have Solved Mummy Mystery." *New York Times*, June 27, 2007.

https://www.nytimes.com/2007/06/27/world/middleeast/27mummy.html.

Wilson, E. B. "The Queen Who Would Be King." *Smithsonian Magazine*, September 2006.

https://www.smithsonianmag.com/history/the-queen-who-would-be-king-130328511.

ZEUS'S PENIS

Aristotle. *Generation of Animals*. Translated by A. L. Peck. Loeb Classical Library 366. Cambridge, MA: Harvard University Press, 1942.

Chrystal, P. *In Bed with the Ancient Greeks*. Stroud, UK: Amberley Publishing, 2016.

Hubbard, T. K. *Homosexuality in Greece and Rome: A Sourcebook of Basic Documents*. Berkeley: University of California Press, 2003.

RESOURCES

Jenkins, I. *Defining Beauty: The Body in Ancient Greek Art*. London:
  British Museum Press, 2015.

McNiven, T. J. "The Unheroic Penis: Otherness Exposed." *Notes in the
  History of Art* 15, no. 1 (1995): 10–16. https://www.jstor.org/stable/23205709.

North, H. F. "The Concept of Sophrosyne in Greek Literary Criticism."
  North Classical Philology 43, no. 1 (1948): 1–17.

CLEOPATRA'S NOSE

Ahmed, E. M., and W. F. Ibrahim. "Hellenistic Heads of Queen Cleopatra VII."
  Journal of Tourism, Hotels and Heritage 1, no. 2, (2020): 30–39. https://sjs.
  journals.ekb.eg/article_125082_35e077628922bdab4ad39c8049716aca.pdf.

Ashton, S. A. "Ptolemaic Royal Sculpture from Egypt: The Greek and Egyptian
  Traditions and Their Interaction." Doctoral dissertation, University of
  London, 1999.

Bianchi, R. S., et. al. *Cleopatra's Egypt: Age of the Ptolemies*. New York:
  The Brooklyn Museum, 1988.

"Cleopatra and Egypt." Humanities Department, Macquarie University.
  http://www.humanities.mq.edu.au/acans/caesar/CivilWars_Cleopatra.htm.

Kleiner, D. E. E. *Cleopatra and Rome*. Cambridge, MA: Harvard University Press,
  2009.

Pascal, B. *Pensées*. London: Penguin Books Ltd., 2003.

Walker, S., and P. Higgs, eds. *Cleopatra of Egypt: From History to Myth*.
  London: British Museum Press, 2001.

Walker, S. "Cleopatra in Pompeii?" *Papers of the British School at Rome*
  76 (2008): 35–46, 345–348.

RESOURCES

TRIỆU THỊ TRINH'S BREASTS

Dasen, V. "Pobaskania: Amulets and Magic in Antiquity." In *The Materiality of Magic*, edited by D. Boschung and J. N. Bremmer, 177–204. Cologne, Germany: Internationales Kolleg Morphomata, 2015.

Gilbert, M. J. "When Heroism Is Not Enough: Three Women Warriors of Vietnam, Their Historians and World History." *World History Connected*, June 2007. https://worldhistoryconnected.press.uillinois.edu/4.3/gilbert.html.

Johns, C. *Sex or Symbol? Erotic Images of Greece and Rome*. London: British Museum Press, 1982.

Jones, D. E. *Women Warriors: A History.* London: Brassey's Military Books, 1997.

Kim, T. T. *Việt Nam sử lược* (*A Brief History of Vietnam*). Hanoi: Nhà xuất bản Văn Học, 2018.

Le, P. H., ed. *Complete Annals of Great Viet*. Hanoi: Khoa học xã hội, 1998.

Marr, D. G. *Vietnamese Tradition on Trial, 1920–1945*. Berkeley: University of California Press, 1984.

Ngọc, H. *Viet Nam: Tradition and Change*. Athens, OH: Ohio University Press, 2016.

Nguyễn, K. V. *Vietnam: A Long History*. Hanoi: Gioi Publishers, 2002.

Silver, C. "Romans Used to Ward Off Sickness with Flying Penis Amulets." *Atlas Obscura*, December 28, 2016.

https://www.atlasobscura.com/articles/romans-used-to-ward-off-sickness-with-flying-penis-amulets.

Taylor, K. W. *The Birth of Vietnam*. Berkeley: University of California Press, 1983.

Williams, C. A. *Roman Homosexuality: Ideologies of Masculinity in Classical Antiquity*. New York: City University of New York, 1999.

ST. CUTHBERT'S FINGERNAILS

Battiscombe, C. F. , ed. *The Relics of Saint Cuthbert*. Oxford: Oxford University Press, 1956.

Bede. *The Life and Miracles of St. Cuthbert, Bishop of Lindesfarne (721)*. https://sourcebooks.fordham.edu/basis/bede-cuthbert.asp.

Biggs, S. J. "A Menagerie of Miracles: The Illustrated Life of St Cuthbert." *Medieval Manuscripts* blog, British Library, January 30, 2013. https://britishlibrary.typepad.co.uk/digitisedmanuscripts/2013/01/a-menagerie-of-miracles-the-illustrated-life-of-st-cuthbert.html.

Boehm, B. D. "Relics and Reliquaries in Medieval Christianity." Department of Medieval Art and The Cloisters, The Metropolitan Museum of Art, 2011. https://www.metmuseum.org/toah/hd/relc/hd_relc.htm.

Colgrave, B., ed. and trans. *Two Lives of Saint Cuthbert: A Life by an Anonymous Monk of Lindisfarne and Bede's Prose Life*. New York: Greenwood Press, 1969.

Cronyn, J. M., and C. V. Horie. *St. Cuthbert's Coffin*. Durham, UK: Dean and Chapter of Durham Cathedral, 1985.

Gayford, M. "Treasures of Heaven: Saints, Relics, and Devotion in Medieval Europe, British Museum." *The Telegraph*, June 10, 2011.

https://www.telegraph.co.uk/culture/art/8565805/Treasures-of-Heaven-Saints-Relics-and-Devotion-in-Medieval-Europe-British-Museum.html.

The Slaves of the Immaculate Heart of Mary. "The Finding of the Tongue of Saint Anthony of Padua (1263)." Catholicism.org, February 15, 2000.

https://catholicism.org/the-finding-of-the-tongue-of-saint-anthony-of-padua-1263.html.

LADY XOC'S TONGUE

Munson, J., V. Amati, M. Collard, and M. J. Macri. "Classic Maya Bloodletting and the Cultural Evolution of Religious Rituals: Quantifying Patterns of Variation in Hieroglyphic Texts." *PLoS One*, September 25, 2014. https://doi.org/10.1371/journal.pone.0107982.

Schele, L., and M. E. Miller. *Blood of Kings: Dynasty and Ritual in Maya Art.* New York: George Braziller, 1992.

Steiger, K. R. *Crosses, Flowers, and Toads: Classic Maya Bloodletting Iconography in Yaxchilan Lintels 24, 25, and 26.* Provo, UT: Brigham Young University, 2010.

AL-MA'ARRI'S EYES

Bosker, M., E. Buringh, and J. L. van Zanden. "From Baghdad to London: Unraveling Urban Development in Europe, the Middle East, and North Africa, 800–1800." *The Review of Economics and Statistics* 95, no. 4 (2013): 1418–1437.

Margoliouth, D. S. "Abu 'l-'Ala al-Ma'arri's Correspondence on Vegetarianism." *Journal of the Royal Asiatic Society* (1902): 289.

Margoliouth, D. S. *Anecdota Oxoniensia: The Letters of Abu 'l-Ala of Ma'arrat.* Oxford, UK: Clarendon Press, 1898.

Rihani, A. *The Luzumiyat of Abu'l-Ala: Selected from His Luzum ma la Yalzam.* New York: James T. White, 1920.

"Syrian Poet Al-Ma'arri: Through the Lens of Disability Studies." Arablit.org, March 24, 2015. https://arablit.org/2015/03/24/syrian-poet-al-maarri-through-the-lens-of-disability-studies.

TIMUR'S (TAMERLANE'S) LEG

De Clavijo, G. *Embassy to Tamerlane, 1403–1406*. London: G. Routledge & Sons, 1928.

Froggatt, P. "The Albinism of Timur, Zal, and Edward The Confessor." *Medical History* 6, no. 4 (1962): 328–342. doi: 10.1017/s0025727300027666.

Gerasimoc, M. M. *The Face Finder*. London: Hutchinsons, 1971.

Manz, B. F. *The Rise and Rule of Tamerlane*. Cambridge, UK: Cambridge University Press, 1989.

Manz, B. F. "Tamerlane's Career and Its Uses." *Journal of World History* 13, no. 1 (2002): 1–25. http://www.jstor.org/stable/20078942.

Quinn, S. A. "Notes on Timurid Legitimacy in Three Safavid Chronicles." *Iranian Studies* 31, no. 2 (2007): 149–158. doi: 10.1080/00210869808701902.

Sela, R. *The Legendary Biographies of Tamerlane: Islam and Heroic Apocrypha in Central Asia*. New York: Cambridge University Press, 2011.

RICHARD III'S BACK

Appleby, J., et al. "The Scoliosis of Richard III, Last Plantagenet King of England: Diagnosis and Clinical Significance." *Lancet* 383, no. 9932 (2014): 19–44. https://doi.org/10.1016/S0140-6736(14)60762-5.

Barras, C. "Teen Growth Spurt Left Richard III with Crooked Spine." *New Scientist*, May 29, 2014. https://www.newscientist.com/article/dn25651-teen-growth-spurt-left-richard-iii-with-crooked-spine.

Chappell, B. "Richard III: Not the Hunchback We Thought He Was?" *The Two Way*, NPR.org, May 30, 2014.

https://www.npr.org/sections/thetwo-way/2014/05/30/317363287/richard-iii-not-the-hunchback-we-thought-he-was.

Cunningham, S. *Richard III: A Royal Enigma*. London: Bloomsbury Academic, 2003.

Lund, M. A. "Richard's Back: Death, Scoliosis and Myth Making." *Medical Humanities* 41 (2015): 89–94.

Metzler, I. *A Social History of Disability in the Middle Ages.* London: Routledge, 2013.

More, T. *The History of King Richard the Thirde (1513), in Workes.* London: John Cawod, John Waly, and Richarde Tottell, 1557.

Rainolde, R. *The Foundacion of Rhetorike.* London: Ihon Kingston, 1563.

Rous, J. "Historia Regum Angliae (1486)." In *Richard III and His Early Historians, 1483–1535*, edited by T. Hearne. Oxford: Clarendon Press, 1975.

Shakespeare, W. "2 Henry VI (1590–91)." In *The Riverside Shakespeare*, 2nd ed., edited by G. Blakemore Evans and J. J. M. Tobin. Boston: Houghton Mifflin, 1997.

Shakespeare, W. "3 Henry VI (1590–91)." In *The Riverside Shakespeare*, 2nd ed., edited by G. Blakemore Evans and J. J. M. Tobin. Boston: Houghton Mifflin, 1997.

Shakespeare, W. "Richard III (1592-93)." In *The Riverside Shakespeare*, 2nd ed., edited by G. Blakemore Evans and J. J. M. Tobin. Boston: Houghton Mifflin, 1997.

Vergil, P. *English History (1512–13).* London: J. B. Nichols and Son, 1844.

MARTIN LUTHER'S BOWELS

BBC News. "Luther's Lavatory Thrills Experts." BBC News, October 22, 2004. http://news.bbc.co.uk/2/hi/europe/3944549.stm.

Leppin, V. *Martin Luther: A Late Medieval Life.* Grand Rapids, MI: Baker Publishing Group, 2017.

Munk, L. "A Little Shit of a Man." *The European Legacy* 5, no. 5 (2010): 725–727. doi: 10.1080/713665526.

Oberman, H. "Teufelsdreck: Eschatology and Scatology in the 'Old' Luther." *The Sixteenth Century Journal* 19, no. 3 (1988): 435–450.

Oberman, H. *Luther: Man Between God and the Devil.* New Haven, CT: Yale University Press, 1989.

Roper, L. *Martin Luther: Renegade and Prophet.* New York: Random House, 2017.

Rupp. E. G. "John Osborne and the Historical Luther." *The Expository Times* 73, no. 5 (1962): 147–151. doi:10.1177/001452466207300505.

Simon, E. *Printed in Utopia: The Renaissance's Radicalism.* Ropley, UK: John Hunt Publishing, 2020.

Skjelver, Danielle Meade. "German Hercules: The Impact of Scatology on the Image of Martin Luther as a Man, 1483-1546." University of Maryland University College. 1-54.

Wetzel, A., ed. *Radicalism and Dissent in the World of Protestant Reform.* Göttingen: Vandenhoeck & Ruprecht, 2017.

ANNE BOLEYN'S HEART

Angell, C. *Heart Burial.* London: Allen and Unwin, 1933.

Bagliani, A. P. "The Corpse in the Middle Ages: The Problem of the Division of the Body." In *The Medieval World*, edited by P. Linehan and J. L. Nelson, 328–330. New York: Routledge, 2001.

Bain, F. E. *Dismemberment in the Medieval and Early Modern English Imaginary: The Performance of Difference.* Kalamazoo, MI: Medieval Institute Publications, 2020.

Brown, E. A. R. "Death and the Human Body in the Late Middle Ages: The Legislation of Boniface VIII on the Division of the Corpse." *Viator* 12 (1981): 221–270.

Foreman, A. "Burying the Body in One Place and the Heart in Another." *Wall Street Journal*, October 31, 2014.

https://www.wsj.com/articles/
burying-the-body-in-one-place-and-the-heart-in-another-1414779035.

Meier, A. "Bury My Heart Apart from Me: The History of Heart Burial." *Atlas Obscura*, February 14, 2014. https://www.atlasobscura.com/articles/heart-burial.

Park, K. "The Life of the Corpse: Division and Dissection in Late Medieval Europe." *Journal of the History of Medicine and Allied Sciences* 50, no. 1 (1995): 111–132. https://doi.org/10.1093/jhmas/50.1.111.

Rebay-Salisbury, K., M. L. Stig Sorensen, and J. Hughes, eds. *Body Parts and Bodies Whole (Studies in Funerary Archaeology)*. Oxford, UK: Oxbow Books, 2010.

Weiss-Krejci, E. "Restless Corpses: 'Secondary Burial' in the Babenberg and Habsburg Dynasties." *Antiquity* 75, no. 290 (2001): 769–780.

CHARLES I'S AND OLIVER CROMWELL'S HEADS

Clymer, L. "Cromwell's Head and Milton's Hair: Corpse Theory in Spectacular Bodies of the Interregnum." *The Eighteenth Century* 40, no. 2 (1999): 91–112.

Meyers, J. "Invitation to a Beheading." *Law and Literature* 25, no. 2 (2013): 268–285.

Preston, P. S. "The Severed Head of Charles I of England Its Use as a Political Stimulus." *Winterthur Portfolio* 6 (1970): 1–13. https://www.journals.uchicago.edu/doi/abs/10.1086/495793?journalCode=wp.

Sauer, E. "Milton and the Stage-Work of Charles I." *Prose Studies* 23, no. 1 (2008): 121–146. doi: 10.1080/01440350008586698.

Skerpan-Wheeler, E. "The First 'Royal': Charles I as Celebrity." *Publications of the Modern Language Association of America* 126, no. 4 (2020): 912–934. doi: 10.1632/pmla.2011.126.4.912.

CHARLES II OF SPAIN'S JAW

Alvarez, G., et al. "The Role of Inbreeding in the Extinction of a European Royal Dynasty." *PloS One* 4, no. 4 (2009): e5174. doi: 10.1371/journal.pone.0005174.

Dominguez Ortiz, A. *The Golden Age of Spain, 1516–1659.* Oxford, UK: Oxford University Press, 1971.

Edwards, J. *The Spain of the Catholic Monarchs, 1474–1520.* New York: Blackwell, 2000.

Parker, G. *Emperor: A New Life of Charles V.* New Haven, CT: Yale University Press, 2019.

Saplakoglu, Y. "Inbreeding Caused the Distinctive 'Habsburg Jaw' of 17th Century Royals That Ruled Europe." *Live Science*, December 2, 2019. https://www.livescience.com/habsburg-jaw-inbreeding.html.

Thompson, E. M., and R. M. Winter. "Another Family with the 'Habsburg Jaw.'" *Journal of Medical Genetics* 25, no. 12 (1988): 838–842. doi: 10.1136/jmg.25.12.838.

Thulin, L. "The Distinctive 'Habsburg Jaw' Was Likely the Result of the Royal Family's Inbreeding." *Smithsonian Magazine*, December 4, 2019.

https://www.smithsonianmag.com/smart-news/distinctive-habsburg-jaw-was-likely-result-royal-familys-inbreeding-180973688.

Yong, E. "How Inbreeding Killed Off a Line of kings." *National Geographic*, April 14, 2009.

https://www.nationalgeographic.com/science/article/how-inbreeding-killed-off-a-line-of-kings.

GEORGE WASHINGTON'S (FAKE) TEETH

Coard, M. "George Washington's Teeth 'Yanked' from Slaves' Mouths." *The Philadelphia Tribune*, February 17, 2020.
https://www.phillytrib.com/commentary/michaelcoard/coard-george-washington-s-teeth-yanked-from-slaves-mouths/article_27b78ce6-dace-563c-a170-34f02626d7e5.html.

Dorr, L. "Presidential False Teeth: The Myth of George Washington's Dentures, Debunked." *Dental Products Report*, June 30, 2015. https://www.dentalproductsreport.com/view/presidential-false-teeth-myth-george-washingtons-dentures-debunked.

Gehred, K. "Did George Washington's False Teeth Come from His Slaves?: A Look at the Evidence, the Responses to That Evidence, and the Limitations of History." *Washington Papers*, October 19, 2016. https://washingtonpapers.org/george-washingtons-false-teeth-come-slaves-look-evidence-responses-evidence-limitations-history.

"George Washington and Teeth from Enslaved People." Washington Library. https://www.mountvernon.org/george-washington/health/washingtons-teeth/george-washington-and-slave-teeth.

"History of Dentures - Invention of Dentures." History of Dentistry, 2021. http://www.historyofdentistry.net/dentistry-history/history-of-dentures.

Thacker, B. "Disease in the Revolutionary War." Washington Library. https://www.mountvernon.org/library/digitalhistory/digital-encyclopedia/article/disease-in-the-revolutionary-war.

Wiencek, H. *An Imperfect God: George Washington, His Slaves, and the Creation of America*. New York: Farrar, Straus and Giroux, 2004.

BENEDICT ARNOLD'S LEG

Brandt, C. *The Man in the Mirror: A Life of Benedict Arnold*. New York: Random House, 1994.

Flexner, J. T. *The Traitor and the Spy: Benedict Arnold and John André*. New York: Harcourt Brace, 1953.

Grant-Costa, P. "Benedict Arnold's Heroic Leg." *Yale Campus Press*, September 17, 2014. https://campuspress.yale.edu/yipp/benedict-arnolds-heroic-leg.

Martin, J. K. *Benedict Arnold: Revolutionary Hero (An American Warrior Reconsidered)*. New York: New York University Press, 1997.

Randall, W. S. *Benedict Arnold: Patriot and Traitor*. New York: William Morrow Inc., 1990.

Seven, J. "Why Did Benedict Arnold Betray America?" *History*, July 17, 2018. https://www.history.com/news/why-did-benedict-arnold-betray-america.

MARAT'S SKIN

Conner, C. D. *Jean Paul Marat: Tribune of the French Revolution*. London: Pluto Press, 2012.

Glover, M. "Great Works: The Death of Marat, by Jacques-Louis David (1793)." *The Independent*, January 3, 2014.

https://www.independent.co.uk/arts-entertainment/art/great-works/great-works-death-marat-jacques-louis-david-1793-9035080.html.

Gombrich, E. H. *The Story of Art*. Oxford, UK: Phaidon, 1978.

Gottschalk, L. R. *Jean Paul Marat: A Study in Radicalism*. Chicago: The University of Chicago Press, 1967.

Jelinek, J. E. "Jean-Paul Marat: The Differential Diagnosis of His Skin Disease." *The American Journal of Dermatopathology* 1, no. 3 (1979): 251–252.

Schama, S., and J. Livesey. *Citizens: A Chronicle of the French Revolution*. London: Royal National Institute of the Blind, 2005.

LORD BYRON'S FOOT

Browne, D. "The Problem of Byron's Lameness." *Proceedings of the Royal Society of Medicine* 53, no. 6 (1960): 440–442. doi: 10.1177/003591576005300615.

Buzwell, G. "Mary Shelley, *Frankenstein* and the Villa Diodati." *Discovering Literature: Romantics & Victorians*, May 15, 2014. https://www.bl.uk/romantics-and-victorians/articles/ mary-shelley-frankenstein-and-the-villa-diodati#.

Hernigou, P., et al. "History of Clubfoot Treatment, Part I: From Manipulation in Antiquity to Splint and Plaster in Renaissance Before Tenotomy." *International Orthopaedics* 41, no. 8 (2017): 1693–1704. doi: 10.1007/s00264-017-3487-1.

MacCarthy, F. *Byron: Life and Legend.* London: John Murray, 2002.

Marchand, L. A. *Byron: A Biography.* Volumes 1 and 2. New York: Knopf, 1957.

Miller, D. S., and E. Davis. "Disabled Authors and Fictional Counterparts." *Clinical Orthopaedics and Related Research* 89 (1972): 76–93.

Mole, T. "Lord Byron and the End of Fame." *International Journal of Cultural Studies* 11, no. 3 (2008): 343–361. https://doi.org/10.1177/1367877908092589.

HARRIET TUBMAN'S BRAIN

Bradford, S. H. *Scenes in the Life of Harriet Tubman.* Auburn, NY: W. J. Moses, 1869.

Clinton, C. *Harriet Tubman: The Road to Freedom.* Boston: Back Bay Books, 2004.

Hobson, J. "Of 'Sound' and 'Unsound' Body and Mind: Reconfiguring the Heroic Portrait of Harriet Tubman." *Frontiers: A Journal of Women Studies* 40, no. 2 (2019): 193–218. doi: 10.5250/fronjwomestud.40.2.0193.

Humez, J. M. "In Search of Harriet Tubman's Spiritual Autobiography." *NWSA Journal* 5, no. 2 (1993): 162–182. http://www.jstor.org/stable/4316258.

RESOURCES

Oertel, K. T. *Harriet Tubman: Slavery, the Civil War, and Civil Rights in the 19th Century*. New York: Routledge, 2016.

Sabourin, V. M., et al. "Head Injury in Heroes of the Civil War and Its Lasting Influence." *Neurosurgical Focus* 41, no. 1 (2016): E4. doi: 10.3171/2016.3.FOCUS1586.

Seaberg, M., and D. Treffert. "Harriet Tubman an Acquired Savant, Says Rain Man's Doctor: Underground Railroad Heroine Had Profound Gifts After a Head Injury." *Psychology Today*, February 1, 2017.

THE BELL FAMILY'S EARS

Booth, K. *The Invention of Miracles: Language, Power, and Alexander Graham Bell's Quest to End Deafness*. New York: Simon & Schuster, 2021.

Bruce, R. V. *Bell: Alexander Graham Bell and the Conquest of Solitude*. Ithaca, NY: Cornell University Press, 1990.

Gorman, M. E., and W. B. Carlson. "Interpreting Invention as a Cognitive Process: The Case of Alexander Graham Bell, Thomas Edison, and the Telephone." *Science, Technology, & Human Values* 15, no. 2 (1990): 131–164. doi: 10.1177/016224399001500201.

Gray, C. *Reluctant Genius: Alexander Graham Bell and the Passion for Invention*. Toronto: HarperCollins, 2007.

Greenwald, B. H. "The Real 'Toll' of A. G. Bell." *Sign Language Studies* 9, no. 3 (2009): 258–265.

Greenwald, B. H., and J. V. Van Cleve. "A Deaf Variety of the Human Race: Historical Memory, Alexander Graham Bell, and Eugenics." *The Journal of the Gilded Age and Progressive Era* 14, no. 1 (2015): 28–48.

Mitchell, S. H. "The Haunting Influence of Alexander Graham Bell." *American Annals of the Deaf* 116, no. 3 (1971): 349–356. http://www.jstor.org/stable/44394260.

"Signing, Alexander Graham Bell and the NAD." *Through Deaf Eyes*, PBS.org. https://www.pbs.org/weta/throughdeafeyes/deaflife/bell_nad.html.

KAISER WILHELM'S ARM

Clark, C. Kaiser Wilhelm II. New York: Routledge, 2013.

Hubbard, Z. S., et al. "Commentary: Brachial Plexus Injury and the Road to World War I." *Neurosurgery* 82, no. 5 (2018): E132–E135. doi: 10.1093/neuros/nyy034.

Jacoby, M. G. "The Birth of Kaiser William II (1859–1941) and His Birth Injury." *Journal of Medical Biography* 16, no. 3 (2008): 178–183. doi: 10.1258/jmb.2007.007030.

Jain, V., et al. "Kaiser Wilhelm Syndrome: Obstetric Trauma or Placental Insult in a Historical Case Mimicking Erb's Palsy." *Medical Hypotheses* 65, no. 1 (2005): 185–191. doi: 10.1016/j.mehy.2004.12.027.

Kohut, T. A. *Wilhelm II and the Germans: A Study in Leadership.* Oxford, UK: Oxford University Press, 1991.

Owen, J. "Kaiser Wilhelm II's Unnatural Love for His Mother 'Led to a Hatred of Britain.'" *The Independent*, November 16, 2013.

MARY MALLON'S GALLBLADDER

Aronson, S. M. "The Civil Rights of Mary Mallon." *Rhode Island Medicine* 78 (1995): 311–312.

Bourdain, A. *Typhoid Mary.* New York: Bloomsbury, 2001.

Dowd, C. *The Irish and the Origins of American Popular Culture.* Oxfordshire, UK: Routledge, 2018.

Leavitt, J. W. "'Typhoid Mary' Strikes Back: Bacteriological Theory and Practice in Early Twentieth-Century Public Health." *Isis* 83, no. 4 (1992): 608–629. http://www.jstor.org/stable/234261.

Marinelli, F., et al. "Mary Mallon (1869–1938) and the History of Typhoid Fever." *Annals of Gastroenterology* 26, no. 2 (2013): 132–134.

RESOURCES

Soper, G. A. "The Curious Career of Typhoid Mary." *Bulletin of the New York Academy of Medicine* 15, no. 10 (1939): 698–712.

Wald, P. "Cultures and Carriers: 'Typhoid Mary' and the Science of Social Control." *Social Text* no. 52/53 (1997): 181–214. doi: 10.2307/466739.

LENIN'S SKIN

Fann, W. E. "Lenin's Embalmers." *The American Journal of Psychiatry* 156, no. 12 (1999): 2006–2007.

Lophukhin, I. M. *Illness, Death, and the Embalming of V. I. Lenin: Truth and Myths.* Moscow: Republic, 1997.

"Preserving Chairman Mao: Embalming a Body to Maintain a Legacy." *The Guardian*, September 11, 2016. https://www.theguardian.com/world/2016/sep/11/preserving-chairman-mao-embalming-a-body-to-maintain-a-legacy.

Yegorov, O. "After Death Do Us Part: How Russian Embalmers Preserve Lenin and His 'Colleagues.'" *Russia Beyond*, November 16, 2017. https://www.rbth.com/history/326748-after-death-do-us-part-russian-art-of-embalming.

Yurchak, A. "Bodies of Lenin: The Hidden Science of Communist Sovereignty." *Representations* 129 (2015): 116–157.

QIU JIN'S FEET

"1907: Qiu Jin, Chinese Feminist and Revolutionary." ExecutedToday.com, July 15, 2011. http://www.executedtoday.com/tag/chiu-chin.

Hagedorn, L. S. and Y. Zhang (Leaf). "China's Progress Toward Gender Equity: From Bound Feet to Boundless Possibilities." PhD dissertation, Iowa State University, 2010.

Hong, F., and J. A. Mangan. "A Martyr for Modernity: Qui Jin—Feminist, Warrior and Revolutionary." *The International Journal of the History of Sport* 18 (2001): 27–54.

Keeling, R. "The Anti-Footbinding Movement, 1872–1922: A Cause for China Rather Than Chinese Women." *Footnotes* 1 (2008): 12–18.

Wang, D. D-W. *A New Literary History of Modern China*. Cambridge, MA: Harvard University Press, 2017.

Wang, P. *Aching for Beauty: Footbinding in China*. New York: Anchor Books, 2000.

Zarrow, P. "He Zhen and Anarcho-Feminism in China." *The Journal of Asian Studies* 47, no. 4 (1988): 796–813.

EINSTEIN'S BRAIN

Altman, L. K. "So, Is This Why Einstein Was So Brilliant?" *New York Times*, June 18, 1999.

Arenn, C. F., et al. "From Brain Collections to Modern Brain Banks: A Historical Perspective." *Alzheimer's & Dementia: Translational Research & Clinical Interventions* 5 (2019): 52–60. https://www.ncbi.nlm.nih.gov/pmc/articles/PMC6365388.

Burrell, B. *Postcards from the Brain Museum: The Improbable Search for Meaning in the Matter of Famous Minds*. New York: Broadway Books, 2005.

Goff, J. "Mussolini's Mysterious Stay at St. Elizabeths." Boundary Stones, WETA.org, July 28, 2015. https://boundarystones.weta.org/2015/07/28/mussolini%E2%80%99s-mysterious-stay-st-elizabeths.

RESOURCES

Hughes, V. "The Tragic Story of How Einstein's Brain Was Stolen and Wasn't Even Special." *National Geographic*, April 21, 2014. https://www.nationalgeographic.com/science/article/the-tragic-story-of-how-einsteins-brain-was-stolen-and-wasnt-even-special#close.

Kremer, W. "The Strange Afterlife of Einstein's Brain." BBC, April 18, 2015. https://www.bbc.com/news/magazine-32354300.

Lepore, F. E. *Finding Einstein's Brain*. New Brunswick, NJ: Rutgers University Press, 2018.

Levy, S. "My Search for Einstein's Brain." *New Jersey Monthly*, August 1, 1978. https://njmonthly.com/articles/historic-jersey/the-search-for-einsteins-brain.

Murray, S. "Who Stole Einstein's Brain?" *MD Magazine*, April 9, 2019. https://www.hcplive.com/view/who-stole-einsteins-brain.

FRIDA KAHLO'S SPINE

Courtney, C. A. "Frida Kahlo's Life of Chronic Pain." Oxford University Press Blog, January 23, 2017. https://blog.oup.com/2017/01/frida-kahlos-life-of-chronic-pain.

Frida Kahlo Foundation. "Frida Kahlo: Biography." www.frida-kahlo-foundation.org.

Fulleylove, R. "Exploring Frida Kahlo's Relationship with Her Body." Google Arts & Culture. https://artsandculture.google.com/story/EQICSfueb1ivJQ?hl=en.

Herrera, H. *Frida: A Biography of Frida Kahlo*. New York: Harper Perennial, 2002.

Luiselli, V. "Frida Kahlo and the Birth of Fridolatry." *The Guardian*, June 11, 2018. https://www.theguardian.com/artanddesign/2018/jun/11/frida-kahlo-fridolatry-artist-myth.

Olds, D. "Frida Isn't Free: The Murky Waters of Creating Crafts with Frida Kahlo's Image and Name." Craft Industry Alliance, October 22, 2019. https://craftindustryalliance.org/frida-isnt-free-the-murky-waters-of-creating-crafts-with-frida-kahlos-image-and-name.

Rosenthal, M. *Diego and Frida: High Drama in Detroit.* Detroit, MI: Detroit Institute of Arts, 2015.

Salisbury, L. W. "Rolling Over in Her Grave: Frida Kahlo's Trademarks and Commodified Legacy." Center for Art Law, August 2, 2019. https://itsartlaw.org/2019/08/02/rolling-over-in-her-grave-frida-kahlos-trademarks-and-commodified-legacy.

Sola-Santiago, F. "Cringeworthy 1932 Newspaper Clip Called Frida Kahlo 'Wife of the Master Mural Painter' Diego Rivera." Remezcla, August 15, 2018. https://remezcla.com/culture/1932-newspaper-clip-called-frida-kahlo-wife-of-the-master-mural-painter-diego-rivera.

ALAN SHEPHARD'S BLADDER

Best, S. L., and K. A. Maciolek. "How Do Astronauts Urinate?" *Urology* 128 (2019): 9–13.

Brueck, H. "From Peeing in a 'Roll-on Cuff' to Pooping into a Bag: A Brief History of How Astronauts Have Gone to the Bathroom in Space for 58 Years." *Business Insider*, July 17, 2019. https://www.businessinsider.com/how-nasa-astronauts-pee-and-poop-in-space-2018-8.

Hollins, H. "Forgotten Hardware: How to Urinate in a Spacesuit." *Advances in Physiology Education* 37 (2013): 123–128.

Maksel, R. "In the Museum: Toilet Training." *Air and Space Magazine*, September 2009.

Thornton, W., H. Whitmore, and W. Lofland. "An Improved Waste Collection System for Space Flight." SAE Technical Paper, 861014, July 14, 1986. https://doi.org/10.4271/861014.

ACKNOWLEDGMENTS

THIS BOOK WAS SUCH A LABOR OF LOVE TO WRITE, and came to life through the help and support of many people. Of course, we'd like to thank our respective families (Alex, Lauren, Micheal, Randy, Sly, Yvonne) for putting up with us and our all-too-frequent dinnertime conversations about various bodily organs, limbs, and other assorted parts ("Did you know that a diseased gallbladder actually . . . ?").

And thanks to our wonderful editor Becca Hunt, who really grokked the book and whose editing was amazingly seamless; our book designer Jacob Covey, whose design was even better than we dared dream; our sensitivity reader Dominique Lear, whose ideas and suggestions were spot-on; our copy editor Mikayla Butchart, for expert cleanup; our eagle-eyed proofreader Karen Levy; and to everyone else at Chronicle for their able assistance. Many thanks as well to our agent Andrea Somberg, and a very special thanks to Wendy Levinson, who took on the project during Andrea's maternity leave, encouraged us with her cheerleading, and found the book a home at Chronicle.

We couldn't have done this addition to our, um, *body* of work (sorry) without all of you!

KATHRYN PETRAS AND **ROSS PETRAS** are a sibling writing team with over 5.3 million copies of their work in print, including "word nerd" books such as the *New York Times* bestseller *You're Saying It Wrong*. They also co-host the popular eponymous podcast on words, grammar, and etymology with NPR's KMUW. They've worked together on projects since their childhood collaborations in Cairo, Egypt, and although they now live several thousand miles apart—in Granada, Spain, and Toronto, Canada, respectively—they are still happily working together today . . . albeit with an occasional time zone issue.

Library of Congress Cataloging-in-Publication Data:

Names: Petras, Kathryn, author. | Petras, Ross, author.

Title: A history of the world through body parts: the stories behind the organs, appendages, digits, and the like attached to (or detached from) famous bodies / Kathryn Petras, Ross Petras.

Description: San Francisco : Chronicle Books, 2022. | Includes bibliographical references.

Identifiers: LCCN 2021043483 | ISBN 9781797202846 (hardback)

Subjects: LCSH: Human body. | Human physiology. | Human body--History.

Classification: LCC QP38 .P48 2022 | DDC 612--dc23/eng/20211004

LC record available at https://lccn.loc.gov/2021043483

Book design by Unflown | Jacob Covey

Manufactured in India

Chronicle Books LLC
680 Second Street
San Francisco, CA 94107

www.chroniclebooks.com

10 9 8 7 6 5 4 3 2 1